D0776815

Deformations of an Elastic Earth

This is Volume 31 in
INTERNATIONAL GEOPHYSICS SERIES
A series of monographs and textbooks
Edited by WILLIAM L. DONN

A complete list of the books in this series appears at the end of this volume.

Deformations of an Elastic Earth

Paolo Lanzano

U. S. Naval Research Laboratory
Washington, D. C.

1982

ACADEMIC PRESS
A Subsidiary of Harcourt Brace Jovanovich, Publishers

New York London
Paris San Diego San Francisco São Paulo Sydney Tokyo Toronto

Library
I.U.P.
Indiana, Pa.
551.1 L297d
c.1

COPYRIGHT © 1982, BY ACADEMIC PRESS, INC.
ALL RIGHTS RESERVED.
NO PART OF THIS PUBLICATION MAY BE REPRODUCED OR TRANSMITTED IN ANY FORM OR BY ANY MEANS, ELECTRONIC OR MECHANICAL, INCLUDING PHOTOCOPY, RECORDING, OR ANY INFORMATION STORAGE AND RETRIEVAL SYSTEM, WITHOUT PERMISSION IN WRITING FROM THE PUBLISHER.

ACADEMIC PRESS, INC.
111 Fifth Avenue, New York, New York 10003

United Kingdom Edition published by
ACADEMIC PRESS, INC. (LONDON) LTD.
24/28 Oval Road, London NW1 7DX

Library of Congress Cataloging in Publication Data

Lanzano, Paolo.
Deformations of an elastic earth.

(International geophysics series)
Bibliography: p.
Includes index.
1. Earth--Figure--Mathematical models. 2. Tides--Mathematical models. 3. Navier-Stokes equations--Numerical solutions. I. Title. II. Series.
QB283.L36 1982 551.1 82-11653
ISBN 0-12-436620-1

PRINTED IN THE UNITED STATES OF AMERICA

82 83 84 85 9 8 7 6 5 4 3 2 1

To Bernadine

Contents

Preface

To study the real Earth from a dynamical point of view, we must take into account (1) its internal constitution, i.e., the Earth's density profile and elastic parameters, (2) its external shape, (3) its gravity field, and (4) various perturbing potentials. Such a study would then be related to the fields of seismology, geodesy, and astronomy.

Let us begin by considering a few basic facts about these four characteristics.

Our knowledge of the interior of the Earth is provided by seismic studies that have been responsible for determining the velocity of propagation of seismic waves and for measuring the spheroidal and toroidal oscillations experienced by the Earth subsequent to seismic events. From this knowledge a rheological profile of the Earth comprising the variation of its density and of the Lamé elastic parameters along various layers can ultimately be determined.

The external shape of the Earth is the geoid, the determination of which can be achieved by the gravimetric or the astrogeodetic method of physical geodesy, or both, and by satellite geodesy.

The Earth's gravity field is the result of two forces: the gravitational attraction exerted by the whole Earth at each point according to Newton's law and the centrifugal force due to the Earth's rotation. The magnitude of the vector field is the intensity of gravity, whereas its direction at a point provides the direction of the vertical at that point.

The most important perturbing potential is the lunisolar potential (also known as the tidal potential), generated by the Moon and the Sun. This potential varies with time because of the relative motion of these two bodies with respect to the Earth and has been the subject of extensive studies in the fields of celestial mechanics and astronomy. Because the real orbit of the Moon about the Earth is rather elaborate, various perturbing terms—the foremost among them the evection term—must be taken into account, the result being that the orbit differs considerably from the standard Keplerian ellipse.

Let us examine next certain consequences of these tidal perturbations.

1. The most important consequence of the lunisolar potential is the ocean tide, by which the free surface of the ocean conforms to a changing field and remains an equipotential surface.
2. Another consequence of the tidal potential is the Earth tide, that is to say, the yielding of the Earth to a perturbing potential because of its elastic properties. This tide was first considered by Darwin in 1883 and was determined to be about a third of the ocean tide (Darwin, 1883, 1907, pp. 1–69). Love, in 1909, introduced special parameters bearing his name that are related to the elastic deformation of a homogeneous spherical Earth. Like the moments of inertia, these Love numbers represent intrinsic properties of the planet and provide estimates for the maximum rigidity of the solid inner core, which seismological studies cannot provide because transverse waves cannot penetrate the liquid outer core.
3. The lunisolar tidal force acts on the spheroidal Earth by exerting a torque about an equatorial axis; the rotating Earth reacts as a gyroscope, which gives rise to an angular motion that can be decomposed into a precession of the Earth's axis about the normal to the ecliptic, augmented by a minor oscillation about the instantaneous axis known as a nutational motion.
4. The liquid outer core within the Earth reacts to oncoming tidal waves having a tesseral harmonic character by producing resonance. Poincaré (1910) first demonstrated that a fluid ellipsoid contained within a rotating mantle exhibits a free nutational mode. The most advanced theory, formulated by Molodenskii (1961), determined this resonance, but it is still far from the final word on this subject (see Melchior, 1978).
5. The dissipation of energy attributable to the tidal motion contributes to the secular retardation of the Earth's rotation.
6. The tidal contribution to the Earth's potential can affect the orbit of an artificial satellite by approximately 50 m and should be taken into account.

Until recently, the modeling of the global ocean tides was based on the numerical integration of the Laplace tidal equations (LTE); however, the simplifying assumptions of a rigid Earth and of a homogeneous incompressible ocean in hydrostatic equilibrium that were used in order to solve those equations seemed to vitiate the results and failed to give close agreement with tidal observations at midocean islands. The problem should be formulated as a boundary-value problem by considering the continental boundary of the ocean basin and the ocean bottom with friction losses; furthermore, if the attraction of the ocean mass is also taken into account the LTE will be transformed into a system of integrodifferential equations.

Earth tides are primarily the results of two effects. One is the yielding of the elastic Earth to the gravitational attraction of the Moon and the Sun; this direct effect goes by the name of "body tide" and is rather well understood because astronomical observations have provided precise information on the forcing function. There is also an indirect effect attributable to the yielding of the Earth to the load of the harmonically varying ocean tide upon the continents, known as load tide. The corresponding forcing function is not well known, however, not only because of our scanty information on the ocean tides, but also because we still do not know the influence played by the structural heterogeneities in the Earth's crust and upper mantle. However, it is evident that load tides and ocean tides are two interrelated phenomena.

The direct and indirect Earth tide are both ultimately derived from the same astronomical input, and consequently they will have the same frequencies. An eventual lag of these tides with respect to the acting potential should give information about the viscosity of the Earth. However, the spatial behavior of these two tides is quite different: the body tide varies smoothly over the Earth, whereas the load tide presents irregularities because of the discontinuity in the forcing function at the coastline and also because of the circulation motion exhibited by the ocean tide around the amphidromes.

Earth tides produce a periodic variation in the direction of the vertical that is reflected in the variation of latitude. This has implications for observational astronomy and for determining orbits of artificial satellites.

Long-period Earth tides perturb the axis of the Earth that has the largest moment of inertia; to conserve angular momentum, the Earth reacts by changing its rotational rate. First predicted by Jeffreys, this phenomenon was later confirmed by the use of atomic standards (see Munk and MacDonald, 1960).

Amplitudes of the Earth tides reach 30–40 cm and cannot be neglected in making lunar and satellite laser measurements that have accuracies good to a few centimeters.

Crustal loading constitutes a phenomenon whose frequencies are intermediate between those of seismic wave transmission, which is an elastic event, and tectonic deformations, which have an anelastic character; thus it might be expected that the study of loading deformations would be useful in determining the threshold of anelasticity for the crust and upper mantle layers.

A better determination of the surface strains produced by the Earth tides will increase the accuracy of long-base interferometer measurements as well as of other instruments that depend on Earth-fixed length standards.

Cubic expansion of the Earth's crust will produce oscillations in water wells and can have a bearing on the ascertaining of the storage capacity and porosity of an aquifer; in general, the tidal variation of underwater fluids can play a factor in the analysis of tectonic processes.

Ocean load effects are responsible in part for the variation of the vertical component of the gravity field (also referred to as tidal gravity), for surface strains (strain tide), and for the deflection of the vertical (tilt tide). It follows that if for a certain region of the globe the Earth model obtainable from seismic evidence is better known than the cotidal charts of the surrounding ocean, it can be expected that accurate tidal gravity measurements on adjacent lands (which can now reach an accuracy of 10 μgal) can be used to improve upon our knowledge of the ocean tide. Moreover, should the reverse be true, the ocean tide can provide some information on the elastic properties of the crust (see Kuo and Jachens, 1977; Melchior *et al.*, 1981).

Considering the variety of problems for which the influence of the tidal potential plays a significant role, it is appropriate that more precise and sophisticated methods be developed to determine both the ocean and Earth tides.

The purpose of this book is to provide a description of the most recent and advanced research work on the elastic deformations of our planet; these deformations will be determined as linear perturbations of the Earth's hydrostatic equilibrium configuration.

The book is primarily intended for scientists and research workers in this particular field. It is not a compendium of formulas or a collection of recently adopted values for geophysical parameters; rather, it emphasizes methods of approach that are carried through by the use of mathematical tools in such a way as to give a rigorous understanding of the phenomena without allowing the mathematical development to become the main theme of the book.

Not all the ongoing research can be encompassed in a volume of this size: the choice of the topics has necessarily been a decision of the author and has been influenced by his research preferences.

The book is divided into six chapters. The first chapter is an introduction to physical geodesy that will acquaint the reader with such basic ideas as geoid, reference ellipsoid, and gravity anomaly; in it are derived the Stokes and Vening-Meinesz formulas used in gravimetric work, and information on their numerical evaluation is also provided.

In Chapter II the hydrostatic equilibrium theory is derived, which gives rise to the Clairaut equation; the approximation is pursued up to third-order terms according to results developed by the author in a sequence of papers published in collaboration with Z. Kopal. The theory presented therein is quite general and can be applied to the study of planetary interiors.

In the same chapter those seismological concepts are discussed that underlie the determination of a realistic density profile by examination of seismological data. Some details on the 1066A Earth model are furnished because of the popularity that this model has attained among researchers in the geophysical field. The 1066A model and older Earth models are used to solve numerically the Clairaut equation and to determine the shape that the Earth should have according to the hydrostatic theory.

In Chapter III a linearized version of the Navier–Stokes equation is obtained that takes into account the elastic properties of the various layers of the Earth's interior. Also a detailed discussion is given of a mathematical model of spheroidal and toroidal deformations that includes the rotational effects and the Earth's spheroidal shape.

Chapter IV is devoted to questions pertaining to the numerical evaluation of the Navier–Stokes equation. First, asymptotic values are established for the load numbers by making use of the Boussinesq theory of the flat plate; next, these asymptotic values are used to evaluate infinite expansions of spherical harmonics that provide the Earth's spheroidal deformations. The second half of the chapter deals with the dynamics of the liquid core and the core–mantle interaction; investigations initiated by Pekeris are followed and rotational effects are ignored. Numerical results due to Pekeris, Longman, Farrell, Varga, and Wahr are mentioned here.

In Chapter V the precessional and nutational motion of a rigid Earth are studied, including numerical results of Woolard. This is meant to be an introduction to the study of the Chandler wobble, which will be undertaken in a future publication. This chapter is concluded with consideration of the Earth and the Moon as components of comparable size in a rigid binary system, and some results of this author are mentioned.

Chapter VI is devoted to the global ocean tides: the Laplace tidal equation is formulated and some of its series solutions from Darwin and Hough are discussed; it concludes with some considerations on the more sophisticated model developed by Zahel and Schwiderski.

In conclusion, the author, who for the last ten years has been associated with the Naval Research Laboratory (NRL), takes this opportunity to express his gratitude to the U.S. Navy for having provided adequate funding to allow him to pursue research activity in the geophysical field. The third-order theory of hydrostatic equilibrium and the mathematical modeling of an elastic rotating Earth described in this book have been accomplished while the author has been in residence at NRL.

Some material on the numerical solution of the Navier–Stokes equation and the mathematical modeling of the Chandler wobble as influenced by the liquid core, was not ready for inclusion in this book. It is hoped that it will appear in a revised edition or, alternatively, will be the basis for a second volume.

Annandale, Virginia PAOLO LANZANO

Chapter I Concepts of Physical Geodesy

1.00 INTRODUCTION

The purpose of this chapter is to provide some information on how the shape of the Earth, that is to say, the geoid, can be determined from basic gravimetric and/or astrogeodetic measurements before we pass on to a detailed study of the theoretical aspects of hydrostatic and elastic deformations.

The geoid is essentially the real shape of Earth, all main effects being included; it does not account, however, for the topographical features of the Earth, and, in a certain way, it is still idealized as an equilibrium surface.

The basic tools of physical geodesy are (1) astronomical measurements to obtain the direction of the plumb line; (2) linear and azimuthal measurements to ascertain distances and angles in a triangulation net; and (3) gravimetrical measurements, which provide rather accurate values for the gravity anomaly, that is to say, for the difference between gravity on the geoid and gravity on a reference surface, which is commonly assumed to be an ellipsoid of revolution.

Two fundamental formulas, the Stokes formula for geoidal undulations and the Vening-Meinesz formula for vertical deflection, are derived, and some computational procedures for their numerical evaluation are outlined. We also cover the variation in coordinates due to a shift in the reference surface.

Since the reference surface is an ellipsoid of revolution, we begin by providing some properties of a general surface of revolution. We then derive the gravity generated by an ellipsoid, which eventually gives rise to the international gravity formula.

No claims of self-sufficiency and/or completeness are made; we do provide, however, the basic steps and the rationale for the derivation of the fundamental equations.

Useful references for the material covered in this chapter are Struik (1950) for the differential geometric aspects of surfaces; and Heiskanen and Moritz (1967), Bomford (1971), and Molodenskii *et al.* (1962) for the basic aspects and principles of physical geodesy. The notation used herein adheres as much as possible to the standard geodetic notation; a few exceptions are made for reasons of consistency with the other chapters of this book.

1.01 DIFFERENTIAL GEOMETRY OF A SURFACE OF REVOLUTION

A surface of revolution is generated by the rigid rotation of a curve about a fixed axis. Without loss of generality, we may assume that the generating curve lies on a plane through the rotational axis; the curve is then known as the meridional curve of the surface, or simply its meridian.

We introduce a cartesian orthogonal frame $(O; x, y, z)$ with O as its origin and having $\mathbf{e}_1$, $\mathbf{e}_2$, and $\mathbf{e}_3$ as the unit vectors along its three coordinate axes x, y, z, respectively. We also consider a meridional curve represented by the

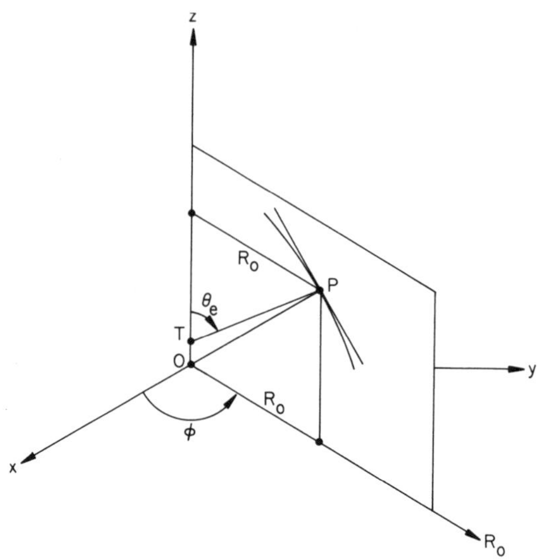

Fig. 1.1 Geometry of a surface of revolution. The segment PT along the normal to the meridian is the radius of normal curvature for the normal section tangent to the parallel of latitude.

equation $R_0(z)$ lying on the $(R_0 z)$-plane. The vector $\overrightarrow{OP}$ to a point P on the curve can be written as a function of the two variables z, ϕ:

$$\mathbf{r}(z, \phi) = R_0(z) \cos \phi \, \mathbf{e}_1 + R_0(z) \sin \phi \, \mathbf{e}_2 + z\mathbf{e}_3, \tag{1}$$

ϕ being the longitude reckoned from the x-axis; see Fig. 1.1. The increment $d\mathbf{r}$ that the radius vector takes on when one moves from P to a neighboring position still on the surface but along the direction defined in space by the increments dz, $d\phi$ is the total differential of $\mathbf{r}(z, \phi)$ considered as a function of the two variables:

$$\begin{aligned} d\mathbf{r} &= \frac{\partial \mathbf{r}}{\partial z} dz + \frac{\partial \mathbf{r}}{\partial \phi} d\phi \\ &= (R_0' \cos \phi \, dz - R_0 \sin \phi \, d\phi)\mathbf{e}_1 \\ &\quad + (R_0' \sin \phi \, dz + R_0 \cos \phi \, d\phi)\mathbf{e}_2 + (dz)\mathbf{e}_3. \end{aligned}$$

Here the prime denotes the derivative with respect to z. The length of this vector is known as the infinitesimal arc length ds of the surface; its square can be written as a quadratic form in dz, $d\phi$:

$$(ds)^2 = d\mathbf{r} \cdot d\mathbf{r} = [1 + (R_0')^2](dz)^2 + R_0^2(d\phi)^2 = E(dz)^2 + G(d\phi)^2, \tag{2}$$

and establishes the metric on the surface. The unit tangent vector in the direction $(dz, d\phi)$ is

$$\mathbf{t} = \frac{d\mathbf{r}}{ds} = \frac{\partial \mathbf{r}}{\partial z} \frac{dz}{ds} + \frac{\partial \mathbf{r}}{\partial \phi} \frac{d\phi}{ds}.$$

Equation (1) is the analytic representation of the surface in terms of the two parameters z, ϕ. It displays the existence of two nets of parametric curves.

The family of curves $\phi =$ const. are the meridians of constant longitude; along any one of them we have $ds = \sqrt{E}\, dz$, so that the differential representation of the meridians is

$$\frac{dz}{ds} = \frac{1}{\sqrt{E}}, \qquad \frac{d\phi}{ds} = 0.$$

The unit tangent vector along a meridian is

$$\mathbf{t}_1 = \frac{\partial \mathbf{r}}{\partial z} \frac{dz}{ds} = \frac{1}{\sqrt{E}} \frac{\partial \mathbf{r}}{\partial z}.$$

The family of curves $z =$ const. is the parallels of constant latitude. Along them $ds = \sqrt{G}\, d\phi$; their representation is

$$\frac{dz}{ds} = 0, \qquad \frac{d\phi}{ds} = \frac{1}{\sqrt{G}}.$$

The unit tangent vector along a parallel is

$$\mathbf{t}_2 = \frac{\partial \mathbf{r}}{\partial \phi}\frac{d\phi}{ds} = \frac{1}{\sqrt{G}}\frac{\partial \mathbf{r}}{\partial \phi}.$$

It is evident that the meridians and the parallels are two sets of orthogonal curves.

Any curve lying on the surface can be represented by a functional relation between the two parameters such as $z = f(\phi)$, or $g(z, \phi) = 0$, or even parametrically as $z = f(u)$, $\phi = g(u)$. In the latter case we have

$$\left(\frac{ds}{du}\right)^2 = E(f')^2 + G(g')^2,$$

where the prime denotes the derivative with respect to the parameter u. The finite length of an arc for such a curve is furnished by the integral

$$s = \int_{u_0}^{u_1} [E(f')^2 + G(g')^2]^{1/2}\, du.$$

Vectors tangent to the surface at P can be decomposed into components tangent to the two parametric curves intersecting at that point. Consider two vectors of length ds_1, ds_2 tangent to the surface and having components $(dz_1, d\phi_1)$, $(dz_2, d\phi_2)$; the cosine of the angle α between them is given by

$$ds_1\, ds_2 \cos\alpha = d\mathbf{r}_1 \cdot d\mathbf{r}_2 = E\, dz_1\, dz_2 + G\, d\phi_1\, d\phi_2.$$

If, in particular, we consider one vector to be tangent to the meridian, i.e., $d\phi_1 = 0$, then $dz_1/ds_1 = 1/\sqrt{E}$ and the cosine of the angle between the meridian and the vector $(dz, d\phi)$ is given by

$$\cos\alpha = E\frac{dz}{ds}\frac{1}{\sqrt{E}} = \sqrt{E}\frac{dz}{ds}. \tag{3}$$

Because of the orthogonality of the parametric curves, it readily follows that

$$\sin\alpha = \sqrt{G}\frac{d\phi}{ds}. \tag{4}$$

The unit vector $\mathbf{n}$ normal to the surface can be written as

$$\begin{aligned}\mathbf{n}(z, \phi) = \mathbf{t}_1 \times \mathbf{t}_2 &= \frac{1}{(EG)^{1/2}}\frac{\partial \mathbf{r}}{\partial z} \times \frac{\partial \mathbf{r}}{\partial \phi}\\ &= [1 + (R_0')^2]^{-1/2}(-\cos\phi\, \mathbf{e}_1 - \sin\phi\, \mathbf{e}_2 + R_0'\mathbf{e}_3).\end{aligned} \tag{5}$$

The components of the vector product have been obtained as the second-order minors extracted from the matrix

$$\begin{bmatrix} R_0' \cos\phi & R_0' \sin\phi & 1 \\ -R_0 \sin\phi & R_0 \cos\phi & 0 \end{bmatrix}.$$

From Eq. (5) it follows that the cosine of the angle θ_e that the normal to the surface (which is also the normal to the meridian) makes with the rotational axis is

$$\cos\theta_e = R_0'[1 + (R_0')^2]^{-1/2}; \tag{6}$$

also that

$$\sin\theta_e = [1 + (R_0')^2]^{-1/2}. \tag{7}$$

The curvature vector **k** for any curve Γ on the surface is $d\mathbf{t}/ds$; the component of this vector along the normal **n** to the surface is called the normal curvature of Γ and is given by

$$\mathbf{k} \cdot \mathbf{n} = \frac{d\mathbf{t}}{ds} \cdot \mathbf{n} = \frac{1}{R}.$$

This normal curvature of Γ can also be interpreted as the curvature for that plane curve γ which is obtained by intersecting the surface with the normal plane passing through the tangent line to Γ.

Since $\mathbf{t} = d\mathbf{r}/ds$, $\mathbf{k} = d^2\mathbf{r}/(ds)^2$ and

$$\frac{1}{R} = \frac{d^2\mathbf{r}}{(ds)^2} \cdot \mathbf{n}.$$

Also, because

$$\begin{aligned} d^2\mathbf{r} = {} & [R_0'' \cos\phi\,(dz)^2 - 2R_0' \sin\phi\,dz\,d\phi - R_0 \cos\phi\,(d\phi)^2]\mathbf{e}_1 \\ & + [R_0'' \sin\phi\,(dz)^2 + 2R_0' \cos\phi\,dz\,d\phi - R_0 \sin\phi\,(d\phi)^2]\mathbf{e}_2, \end{aligned}$$

we get, by using Eq. (5),

$$\begin{aligned} d^2\mathbf{r} \cdot \mathbf{n} &= -R_0''[1 + (R_0')^2]^{-1/2}(dz)^2 + R_0[1 + (R_0')^2]^{-1/2}(d\phi)^2 \\ &= L(dz)^2 + N(d\phi)^2, \end{aligned}$$

leading finally to

$$\frac{1}{R} = \frac{L(dz)^2 + N(d\phi)^2}{E(dz)^2 + G(d\phi)^2}. \tag{8}$$

A study of the variation of the above function reveals that it has only one maximum and only one minimum, occurring at $dz = 0$ and $d\phi = 0$ or vice versa according to the sign of the expression $1 + (R_0')^2 + R_0 R_0''$. These extreme values are at 90° from each other; they occur for the meridian $d\phi = 0$, giving rise to

$$1/R_1 = L/E = -R_0''[1 + (R_0')^2]^{-3/2}, \tag{9}$$

and for the normal section tangent to the parallel $dz = 0$, giving rise to

$$1/R_2 = N/G = R_0^{-1}[1 + (R_0')^2]^{-1/2}. \tag{10}$$

R_1 and R_2 are known as the principal radii of normal curvature and their reciprocals as the principal normal curvatures.

Using Eqs. (3), (4), (9), and (10), we can rewrite Eq. (8) as

$$\frac{1}{R(\alpha)} = L\left(\frac{dz}{ds}\right)^2 + N\left(\frac{d\phi}{ds}\right)^2 = \frac{L}{E}\cos^2\alpha + \frac{N}{G}\sin^2\alpha$$
$$= \frac{1}{R_1}\cos^2\alpha + \frac{1}{R_2}\sin^2\alpha. \tag{11}$$

This is known as the Euler formula for the normal curvature at a certain point and displays the variation of the normal curvature as a function of the azimuthal angle that the normal section forms with the meridian. This variation is bounded between the maximum and the minimum values unless $R_1 = R_2$, in which case the curvature is independent of the azimuth, as it is in the case of the sphere.

Using the Euler formula we can write

$$\frac{1}{R(\alpha)} + \frac{1}{R[\alpha + (\pi/2)]} = \frac{1}{R_1}\cos^2\alpha + \frac{1}{R_2}\sin^2\alpha + \frac{1}{R_1}\sin^2\alpha + \frac{1}{R_2}\cos^2\alpha$$
$$= \frac{1}{R_1} + \frac{1}{R_2}. \tag{12}$$

This shows that the sum of the normal curvatures for two sections perpendicular to each other is independent of the orientation, that is to say, is a function of position alone; it is known as the mean curvature of the surface at that point.

Before bringing this introductory section on surfaces of revolution to a close, we shall make two other remarks pertaining to the interpretation of the principal radii of normal curvature which will be useful in the sequel.

1. From Eq. (10) we have

$$R_0 = R_2[1 + (R_0')^2]^{-1/2} = R_2 \sin\theta_e, \tag{13}$$

where, according to Eq. (7), θ_e is the angle that the normal to the meridian makes with the rotational axis. This shows that the principal radius of

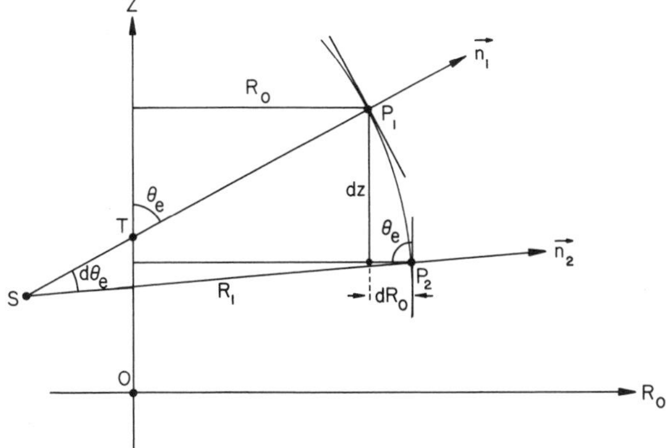

Fig. 1.2 Increment of the coordinates expressed in terms of the radius of curvature of the meridian and the angle between the normals.

normal curvature R_2 is the length PT along the normal **n** starting at the point P and terminating on the axis of rotation; see Fig. 1.1.

2. Since R_1 is the radius of curvature of the meridian, we can see from Fig. 1.2 that we must have $dR_0 = (R_1\, d\theta_e)\cos\theta_e$, so that

$$dR_0/d\theta_e = R_1 \cos\theta_e. \tag{14}$$

Also, from $dz = (R_1\, d\theta_e)\sin\theta_e$, we have $dz/\sin\theta_e = R_1\, d\theta_e$, leading to

$$R_1^2(d\theta_e)^2 = [1 + (R_0')^2](dz)^2 = E(dz)^2,$$

and this provides another representation for the arc length according to

$$(ds)^2 = E(dz)^2 + G(d\phi)^2 = R_1^2(d\theta_e)^2 + R_0^2(d\phi)^2. \tag{15}$$

1.02 SOME PROPERTIES OF THE ELLIPSE. DEFINITION OF ELLIPSOIDAL COORDINATES

The fundamental reference surface in physical geodesy is the ellipsoid of revolution, which is generated by the rigid rotation of an ellipse about a fixed axis. It is appropriate therefore that we recall a few elementary properties of the ellipse. From the equation of the ellipse

$$\frac{R_0^2}{a^2} + \frac{z^2}{b^2} = 1$$

in the $(R_0 z)$-plane with $a > b$, we obtain the parametric representation

$$R_0 = a\sin\theta_r, \qquad z = b\cos\theta_r,$$

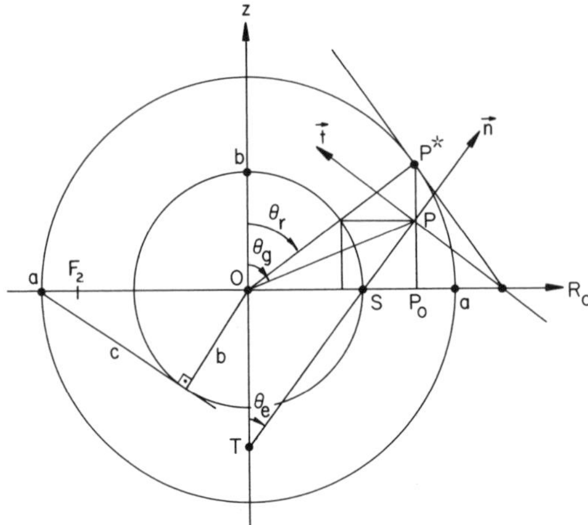

Fig. 1.3 Definition of fundamental ellipsoidal parameters. Pointwise construction for ellipse of given semiaxes (a, b).

and this will allow us to verify the pointwise construction of the ellipse by means of two concentric circles of radii a and b; see Fig. 1.3.

The affine transformation of coordinates

$$R_0 = R_0^*, \qquad z = (b/a)z^*$$

takes the ellipse (a, b) into the concentric circle of radius a. Corresponding points, P^* on the circle and P on the ellipse, have the same abscissa and their ordinates are in the ratio $P_0P^*/P_0P = a/b$. The locus of self-corresponding points according to this affine transformation (i.e., the locus of those points that coincide with their corresponding point) is $z = 0$, i.e., the R_0-axis. Thus the points of intersection of circle and ellipse must be on the R_0-axis. By the same token, the tangent to the circle and the tangent to the ellipse at corresponding points must intersect at a point which must be self-corresponding and consequently located on the R_0-axis. This property allows a simple geometric construction of the tangent to the ellipse and also of its normal; the latter makes the angle θ_e with the rotational axis. The central angle θ_r is called the reduced colatitude of P^* (on the circle), and the central angle θ_g is called the geocentric colatitude of P (on the ellipse). The two foci F_1 and F_2 have the same distance c from the center O, so that $c^2 = a^2 - b^2$; see Fig. 1.3.

From analytic geometry, we calculate the slope that the tangent to the ellipse at one of its points (R_0, z) makes with the R_0-axis to be $-(b/a)^2(R_0/z)$; the slope of its normal, being the negative reciprocal of the above quantity, is then $\cot\theta_e = (a/b)^2(z/R_0)$. It follows that

$$\sin\theta_e = b^2R_0(a^4z^2 + b^4R_0^2)^{-1/2}.$$

We can eliminate the z using the information that the point in question belongs to the ellipse $b^2R_0^2 + a^2z^2 = a^2b^2$. We easily obtain

$$\sin\theta_e = bR_0(a^4 - c^2R_0^2)^{-1/2}.$$

This can be solved to obtain

$$R_0 = a^2\sin\theta_e(b^2 + c^2\sin^2\theta_e)^{-1/2} \tag{16}$$

as the expression of the radius of the circle of latitude in terms of the ellipsoidal colatitude θ_e.

Using the above formula and recalling Eqs. (13) and (14), we can evaluate the principal radii of normal curvature for the ellipsoid of revolution whose minor axis is b:

$$\begin{aligned} R_1(b,\theta_e) &= \frac{1}{\cos\theta_e}\frac{dR_0}{d\theta_e} = a^2b^2(b^2 + c^2\sin^2\theta_e)^{-3/2} \\ &= a(1 - e_1^2)(1 - e_1^2\cos^2\theta_e)^{-3/2} \end{aligned} \tag{17}$$

and

$$\begin{aligned} R_2(b,\theta_e) &= \frac{R_0}{\sin\theta_e} = a^2(b^2 + c^2\sin^2\theta_e)^{-1/2} \\ &= a(1 - e_1^2\cos^2\theta_e)^{-1/2}. \end{aligned} \tag{18}$$

Here $e_1 = c/a$ is the first eccentricity of the ellipse. The ratio

$$R_2/R_1 = (1 - e_1^2\cos^2\theta_e)/(1 - e_1^2)$$

will always be >1, i.e., $R_2 > R_1$, unless we are at the pole $\theta_e = 0$, in which case $R_2 = R_1$. The center C of curvature for the meridian (ellipse) will then be located between P and T (see Fig. 1.3) since

$$R_2 = \overrightarrow{PT}.$$

The minimum value for both radii of curvature is reached at the equator, $\theta_e = \pi/2$; their maximum value at $\theta_e = 0$. From analytic geometry, we can evaluate $\overrightarrow{OS} = R_0(c/a)^2$ and $\overrightarrow{OT} = -z(c/b)^2$, which appear in Fig. 1.3, so that we can write

$$\tan\theta_e = \left|\frac{OS}{OT}\right| = \frac{b^2}{a^2}\frac{R_0}{z} = \frac{b^2}{a^2}\tan\theta_g,$$

whence

$$\tan\theta_e/\tan\theta_g = b^2/a^2. \tag{19}$$

We can also verify that

$$\frac{\tan\theta_g}{\tan\theta_r} = \frac{\cot\theta_r}{\cot\theta_g} = \frac{P_0P^*}{P_0P} = \frac{a}{b}. \tag{20}$$

Hence it follows that

$$\tan\theta_e/\tan\theta_r = b/a. \tag{21}$$

Thus a simple reduction of the three colatitudes is easily available, and it shows that

$$\theta_g > \theta_r > \theta_e.$$

Let us now consider a family of ellipsoids of revolution having the same focal distance $\overrightarrow{OF}_1 = c = \text{const.}$; the varying minor axis will be denoted by the parameter u; the varying major axis is then $a = (u^2 + c^2)^{1/2}$. The cartesian coordinates of any point in space can then be expressed by means of the ellipsoid of the family passing through it:

$$\begin{aligned} x &= (u^2 + c^2)^{1/2} \sin\theta_r \cos\phi, \\ y &= (u^2 + c^2)^{1/2} \sin\theta_r \sin\phi, \\ z &= u\cos\theta_r, \end{aligned} \tag{22}$$

where θ_r is the reduced colatitude of the point (see Fig. 1.3) and ϕ is its longitude reckoned from the x-axis. The distance of P from the origin is then given by

$$r^2 = (OP)^2 = x^2 + y^2 + z^2 = u^2 + c^2 \sin^2\theta_r. \tag{23}$$

The three fundamental coordinate surfaces passing through any given point are as follows:

1. The surfaces $u = \text{const.}$ are ellipsoids of revolution; this can be seen by first eliminating the ϕ-coordinate, which yields

$$x^2 + y^2 = (u^2 + c^2)\sin^2\theta_r, \qquad z^2 = u^2\cos^2\theta_r,$$

then the θ_r, and this ultimately gives

$$\frac{x^2 + y^2}{u^2 + c^2} + \frac{z^2}{u^2} = 1.$$

2. The surfaces $\theta_r = \text{const.}$ are, on the other hand, hyperboloids of revolution of one sheet; this can be seen first by eliminating ϕ:

$$x^2 + y^2 = \left(\frac{u^2}{c^2} + 1\right)c^2\sin^2\theta_r, \qquad u^2 = \frac{z^2}{\cos^2\theta_r};$$

elimination of u gives

$$x^2 + y^2 = \left(\frac{z^2}{c^2 \cos^2 \theta_r} + 1\right) c^2 \sin^2 \theta_r;$$

this can be written as

$$\frac{x^2 + y^2}{c^2 \sin^2 \theta_r} - \frac{z^2}{c^2 \cos^2 \theta_r} = 1,$$

which is recognized as a hyperboloid of one sheet.

3. Finally, the family of surfaces $\phi =$ const. is obtained by eliminating θ_r and u; it gives rise to

$$y/x = \tan \phi,$$

the family of meridional planes.

Equations (22) represent a transformation of coordinates from a cartesian frame to a frame of confocal quadrics. The square of the element of arc in this system is a quadratic form in du, $d\theta_r$, $d\phi$, which can be written $(ds)^2 = g_{ij}\, dx^i\, dx^j$, where the summation convention applies to repeated indices and the metric tensor has components g_{ij} obtainable from the derivatives of Eq. (22).

If we identify $x^1 = u$, $x^2 = \theta_r$, $x^3 = \phi$, we can easily verify that

$$\begin{aligned} g_{11} &= \frac{u^2 + c^2 \cos^2 \theta_r}{u^2 + c^2}, \\ g_{22} &= u^2 + c^2 \cos^2 \theta_r, \\ g_{33} &= (u^2 + c^2) \sin^2 \theta_r. \end{aligned} \tag{24}$$

Also,

$$g_{12} = g_{13} = g_{23} = 0,$$

which shows that the fundamental surfaces are orthogonal to each other.

1.03 POTENTIAL OF THE ELLIPSOID EXPRESSED IN TERMS OF ELLIPSOIDAL COORDINATES

Let us think of the given ellipsoid of revolution E_0, of mass m_e, as the element $u = b$ within the system of confocal quadrics which we have previously discussed.

The gravitational potential, being a harmonic function outside the ellipsoid, must satisfy the Laplace equation. If use is made of ellipsoidal coordinates, one can then ascertain (see, e.g., Heiskanen and Moritz, 1967, p. 43) that the potential can be represented in terms of the Legendre functions of the second kind of imaginary argument. The gravity potential can be written as

$$\sum_{n=0}^{\infty} A_n \frac{Q_n(iu/c)}{Q_n(ib/c)} P_n(\cos\theta_r) + \tfrac{1}{2}\omega^2(u^2 + c^2)\sin^2\theta_r. \tag{25}$$

Here i is the imaginary unit satisfying the relation $i^2 + 1 = 0$, and the Q are the Legendre functions of the second kind. For the complex argument $z = x + iy$, they are defined as follows (see, e.g., Hobson, 1955, p. 70):

$$Q_n(z) = \tfrac{1}{2}P_n(z)\ln\left(\frac{z+1}{z-1}\right) - \sum_{k=1}^{n} \frac{1}{k} P_{k-1}(z)P_{n-k}(z) \tag{26}$$

for $n \geqslant 1$, and

$$Q_0(z) = \frac{1}{2}\ln\left(\frac{z+1}{z-1}\right) = \coth^{-1} z. \tag{27}$$

The ellipsoid E_0 is the surface of the system corresponding to $u = b$ and is also an equipotential surface of the field; it must be true then that for $u = b$ the above expression must assume the same constant value, say U_0, whatever the value of θ_r:

$$\sum_{n=0}^{\infty} A_n P_n(\cos\theta_r) + \tfrac{1}{2}\omega^2 a^2 \sin^2\theta_r \equiv U_0.$$

This means that the coefficients of every Legendre polynomial P_n must vanish. Since

$$\tfrac{1}{2}\sin^2\theta = \tfrac{1}{3}[1 - P_2(\cos\theta)],$$

we get

$$A_0 + \tfrac{1}{3}\omega^2 a^2 - U_0 = 0; \qquad A_2 - \tfrac{1}{3}\omega^2 a^2 = 0;$$
$$A_1 = 0; \qquad A_n = 0, \qquad n \geqslant 3;$$

leading to

$$\begin{aligned} A_0 = U_0 - \tfrac{1}{3}a^2\omega^2; &\qquad A_2 = \tfrac{1}{3}\omega^2 a^2; \\ A_1 = 0; \qquad A_n = 0, &\qquad n \geqslant 3. \end{aligned} \tag{28}$$

The gravity potential becomes

$$A_0 \frac{Q_0(iu/c)}{Q_0(ib/c)} + A_2 \frac{Q_2(iu/c)}{Q_2(ib/c)} P_2(\cos\theta_r) + \tfrac{1}{2}\omega^2(u^2 + c^2)\sin^2\theta_r, \tag{29}$$

where A_0 and A_2 are given by Eq. (28) and, according to Eqs. (26) and (27),

$$\begin{aligned} Q_0(iu/c) &= \coth^{-1}(iu/c) = -i\tan^{-1}(c/u), \\ Q_2\left(\frac{iu}{c}\right) &= \frac{i}{2}\left[\left(1 + 3\frac{u^2}{c^2}\right)\tan^{-1}\left(\frac{c}{u}\right) - 3\frac{u}{c}\right] \equiv \frac{i}{2} Q\left(\frac{c}{u}\right). \end{aligned} \tag{30}$$

It is appropriate at this point to establish the following series expansions valid for large values of u:

$$\begin{aligned} \tan^{-1}\left(\frac{c}{u}\right) &= \sum_{n=0}^{\infty} \frac{(-1)^n}{(2n+1)}\left(\frac{c}{u}\right)^{2n+1}, \\ Q\left(\frac{c}{u}\right) &= \sum_{n=0}^{\infty} (-1)^{n+1} \frac{4n}{(2n+1)(2n+3)}\left(\frac{c}{u}\right)^{2n+1}. \end{aligned} \tag{31}$$

Consider now the asymptotic expansion, for large values of u, of the gravitational portion of the above potential. This implies large values of r because, due to Eq. (23),

$$\frac{c}{u} = \frac{c}{r}\left[1 - \frac{c^2}{r^2}\sin^2\theta_r\right]^{-1/2} = \frac{c}{r} + O(-3).$$

Since the function $\tan^{-1}(c/u)$ begins with terms in c/u and the function $Q(c/u)$ begins with terms in $(c/u)^3$ [see Eq. (31)], we realize that the coefficient of $Q_0(iu/c)$ in Eq. (29) must be such that

$$(U_0 - \tfrac{1}{3}\omega^2 a^2)/\tan^{-1}(c/b) = Gm_e/c,$$

where G is the universal gravitational constant. This is so because the leading term in Eq. (29) should generate the leading term Gm_e/r, which is characteristic of the potential function in spherical coordinates.

Taking all these considerations into account, we can finally write the gravity potential as

$$\begin{aligned} U(b; u, \theta_r) = {} & \frac{Gm_e}{c}\tan^{-1}\left(\frac{c}{u}\right) \\ & + \tfrac{1}{3}\omega^2 a^2 \frac{Q(c/u)}{Q(c/b)} P_2(\cos\theta_r) + \tfrac{1}{2}\omega^2(u^2 + c^2)\sin^2\theta_r, \end{aligned} \tag{32}$$

where both the arctangent function and the Q-function can be expressed as infinite power series in $1/u$ according to Eq. (31).

1.04 NORMAL GRAVITY

The gradient of the aforementioned potential is the gravity γ of the ellipsoid. To calculate the components of the gravity field tangent to each of the parametric lines, one must take the derivative of $U(b; u, \theta_r)$ as given by Eq. (32) with respect to the arc length along each of the parametric curves, that is to say, when only one of the three variables (u, θ_r, ϕ) is allowed to vary and the other two are kept constant.

From the general expression of the $(ds)^2$ in ellipsoidal coordinates [see Eq. (24)], we can write

$$(ds)_u = \sqrt{g_{11}}\, du, \qquad (ds)_\theta = \sqrt{g_{22}}\, d\theta_r, \qquad (ds)_\phi = \sqrt{g_{33}}\, d\phi,$$

where the subscripts denote the variables that are allowed to vary. Thus

$$\gamma_\phi(u, \theta_r) = \frac{1}{\sqrt{g_{33}}} \frac{\partial U}{\partial \phi} \equiv 0$$

since U does not contain the ϕ-variable due to rotational symmetry;

$$\gamma_\theta(u, \theta_r) = \frac{1}{\sqrt{g_{22}}} \frac{\partial U}{\partial \theta_r}$$

can be shown to vanish on the surface of the ellipsoid, i.e.,

$$\gamma_\theta(b, \theta_r) \equiv 0;$$

and finally, the normal gravity component is

$$\gamma_u(u, \theta_r) = \frac{1}{\sqrt{g_{11}}} \frac{\partial U}{\partial u},$$

where $g_{11} = (u^2 + c^2 \cos^2 \theta_r)/(u^2 + c^2)$ according to Eq. (24). Since this turns out to be the only nonvanishing component of the gravity field at the surface of the ellipsoid, we shall omit the subscript u and write henceforth γ instead of γ_u for the normal component without any fear of misunderstanding.

By making use of the trigonometric representation of P_2 [see Eq. (34) in Chapter II], and of other trigonometric identities, we can write the normal gravity on the ellipsoid ($u = b$) as

$$-\gamma(b, \theta_r) = \frac{Gm_e/a}{(a^2 \cos^2 \theta_r + b^2 \sin^2 \theta_r)^{1/2}} \times \left\{ \left[1 + q \frac{ca^2}{b^3} \frac{Q'(c/b)}{Q(c/b)} \right] \cos^2 \theta_r + \left[1 - 3q - \frac{1}{2} q \frac{ca^2}{b^3} \frac{Q'(c/b)}{Q(c/b)} \right] \sin^2 \theta_r \right\}, \tag{33}$$

where $q = \omega^2 a^2 b / 3Gm_e$ is the so-called rotational parameter (a dimensionless quantity) and the primes denote the derivatives of the series $Q(c/u)$ with respect to the quantity c/u; this causes Q' also to be a dimensionless quantity. The negative sign shows that the normal component γ is directed toward the interior of the ellipsoid.

If we evaluate the normal component of gravity at the pole ($\theta_r = 0$) and at the equator ($\theta_r = \pi/2$), we can rewrite the previous formula as

$$\gamma(b, \theta_r) = \frac{a\gamma(b, 0) \cos^2 \theta_r + b\gamma(b, \pi/2) \sin^2 \theta_r}{(a^2 \cos^2 \theta_r + b^2 \sin^2 \theta_r)^{1/2}}, \tag{34}$$

which displays the composition of the gravity at any latitude in terms of the gravity at the pole and at the equator. There is a similar formula in terms of the ellipsoidal colatitude θ_e obtainable from the preceeding one by using Eq. (21). We get a formula due to Somigliana,

$$\gamma(b, \theta_e) = \frac{a\gamma(b, \pi/2) \sin^2 \theta_e + b\gamma(b, 0) \cos^2 \theta_e}{(a^2 \sin^2 \theta_e + b^2 \cos^2 \theta_e)^{1/2}}, \tag{35}$$

which happens to be more useful than the previous one because in practice the angle θ_e is more easily assessed than θ_r. Normal gravity at a small elevation h above the ellipsoid can be approximated from the normal gravity at the ellipsoid surface by the series expansion

$$\gamma(b + h, \theta_e) = \gamma(b, \theta_e) + \frac{\partial \gamma(b, \theta_e)}{\partial h} h + \cdots. \tag{36}$$

The differential coefficient can be obtained in terms of the mean curvature M of the ellipsoid. The most expeditious way to ascertain this dependence is to use cartesian coordinates. The result we obtain here is more general than what is required in this specific instance and refers to a general vector field.

Let $P(x, y, z) = \text{const.}$ be an equipotential surface for the geopotential represented in cartesian coordinates. At a point $Q(x_0, y_0, z_0)$ on the surface, consider its tangent plane as the (xy)-plane and the normal to the surface as the z-axis. We wish to determine the curvature $K(x, z)$ of the curve which is obtained by intersecting the equipotential surface with the (xz)-plane and is represented implicitly by $P(x, y_0, z) = \text{const.}$ If we differentiate this equation implicitly twice with respect to x, considering $z(x)$ defined by it, we easily see that

$$P_{xx} + 2P_{xz}\frac{dz}{dx} + P_{zz}\left(\frac{dz}{dx}\right)^2 + P_z\frac{d^2z}{dx^2} = 0,$$

where the subscripts stand for partial differentiation with respect to the same variables. In view of the fact that at Q we have $dz/dx = 0$, we see that the curvature is given by

$$K(x, z) = \frac{d^2z}{dx^2} = -\frac{P_{xx}}{P_z} = \frac{P_{xx}}{g};$$

the last step is due to the fact that the negative gradient of the equipotential is the gravity field.

By interchanging the x- with the y-axis, we get

$$K(y, z) = \frac{d^2z}{dy^2} = \frac{P_{yy}}{g},$$

which represents the curvature of the curve obtained as intersection of the surface with the (yz)-plane. We know that the sum of the curvatures of any two normal sections at right angles to each other does not depend on the particular pair and is called the mean curvature M of the surface at the point. Thus

$$K(x, z) + K(y, z) = M = \frac{1}{g}(P_{xx} + P_{yy}) = \frac{1}{R_1} + \frac{1}{R_2}$$

is a function of the point $Q(x_0, y_0, z_0)$.

Let us now consider the Poisson equation satisfied by the equipotential surface

$$\nabla^2 P \equiv P_{xx} + P_{yy} + P_{zz} = -4\pi G\rho_0 + 2\omega^2,$$

where $\rho_0(x, y, z)$ is the mass density at the location under consideration (equals zero if we deal with the exterior potential); we get

$$gM + P_{zz} = -4\pi G\rho_0 + 2\omega^2,$$

or

$$\frac{\partial g}{\partial h} = gM + 4\pi G\rho_0 - 2\omega^2, \tag{37}$$

which expresses the vertical gradient of gravity and goes by the name of the Bruns formula. Reverting to our problem on the ellipsoid, by assuming zero density at the outer surface we can write

$$\frac{\partial \gamma(b, \theta_e)}{\partial h} = -\gamma(b, \theta_e)\left(\frac{1}{|R_1|} + \frac{1}{|R_2|}\right) - 2\omega^2, \tag{38}$$

where the expressions for the principal radii of normal curvature R_1 and R_2 are given by Eqs. (17) and (18). The mean curvature is taken to be negative for the ellipsoid.

1.05 SOME SERIES EXPANSIONS AND THE GRAVITY FORMULA

In studying the ellipsoid, we introduce a number of nondimensional small parameters, whose values are less than one:

$$\begin{aligned} e_1 &= c/a && \text{(first eccentricity)},\\ e_2 &= c/b && \text{(second eccentricity)},\\ f &= (a-b)/a && \text{(flattening)},\\ q &= \omega^2 a^2 b/3Gm_e && \text{(rotational parameter)}; \end{aligned} \tag{39}$$

and seek to establish a few series expansions for the gravity of the ellipsoid, expressed in terms of e_2 and f, which have practical utility.

Gravity at the pole is obtained from Eq. (33) for $\theta_e = 0$ and can be expressed as

$$\begin{aligned} -\gamma(b, 0) &= \frac{Gm_e}{a^2}\left[1 + q\frac{ca^2}{b^3}\frac{Q'(c/b)}{Q(c/b)}\right]\\ &= \frac{Gm_e}{a^2}\left(1 + 3q + \tfrac{18}{7} fq\right), \end{aligned} \tag{40}$$

where use has been made of the series expansion of $Q(c/u)$ given by Eq. (31), the relation $(a/b)^2 = 1 + e_2^2$, and the expansion

$$(c/b)^2 = e_2^2 = (1-f)^{-2} - 1 = 2f + 3f^2 + 4f^3 + \cdots.$$

Gravity at the equator is obtainable from Eq. (33) for $\theta_e = \pi/2$:

$$-\gamma(b;\pi/2) = \frac{Gm_e}{ab}\left[1 - 3q - \tfrac{1}{2}q\,\frac{ca^2}{b^3}\frac{Q'(c/b)}{Q(c/b)}\right]$$

$$= \frac{Gm_e}{ab}\left(1 - \tfrac{9}{2}q - \tfrac{9}{7}fq\right), \tag{41}$$

where use has been made here also of the relation

$$(b/a)^2 = (1 - f)^2 = 1 - 2f + f^2.$$

Let us now rewrite the Somigliana formula [Eq. (35)] in the form

$$\gamma(b;\theta_e) = \gamma\left(b;\frac{\pi}{2}\right)\frac{1 + A\cos^2\theta_e}{(1 - B\cos^2\theta_e)^{1/2}}, \tag{42}$$

where

$$A = \frac{b\gamma(b;0) - a\gamma(b;\pi/2)}{a\gamma(b;\pi/2)} = -2f + \tfrac{15}{2}q + f^2 - \tfrac{78}{7}fq + \tfrac{135}{4}q^2 + \cdots,$$

$$B = 1 - (b/a)^2 = 2f - f^2$$

have been evaluated by using the already mentioned expansions, and where we have limited ourselves to terms of the second degree in f and q. If one now uses the well-known expansion

$$(1 - x)^{-1/2} = 1 + \tfrac{1}{2}x + \tfrac{3}{8}x^2 + \cdots,$$

then Eq. (42) can be put into the form

$$\gamma(b;\theta_e) = \gamma(b;\pi/2)(1 + f_2\cos^2\theta_e + f_4\cos^4\theta_e), \tag{43}$$

and the coefficients f_2 and f_4 can be seen to be of the form

$$\begin{aligned} f_2 &= -f + \tfrac{15}{2}q + \tfrac{1}{2}f^2 - \tfrac{78}{7}fq + \tfrac{135}{4}q^2 + \cdots, \\ f_4 &= -\tfrac{1}{2}f^2 + \tfrac{15}{2}fq + \cdots, \end{aligned} \tag{44}$$

when evaluated up to the second-order terms in f and q. Equation (43) is the most important of the expansions and is known as the gravity formula when the coefficients f_2, f_4 are specified. The most commonly used specification goes by the name of Geodetic Reference System (GRS67) and was adopted at the 1971 Moscow meeting of the IAG. The parameters are as follows:

$$\begin{aligned} \gamma(b;\pi/2) &= 978.0318456 \quad &&\text{gal, including the atmosphere,} \\ f_2 &= 0.005278895 \quad &&\text{(nondimensional),} \\ f_4 &= 0.000023462 \quad &&\text{(nondimensional).} \end{aligned} \tag{45}$$

It is based on the revised value of the Potsdam gravity reference datum, set at $g = 981.260$ gal, and depends also on the adopted values of the following four constants:

$$
\begin{aligned}
a &= 6{,}378{,}160 && \text{ms,} \\
Gm &= 3.98603 \cdot 10^{20} && \text{cm}^3/\text{sec}^2 \quad \text{(including the atmosphere),} \\
K_2 &= 0.0010827, && \text{nondimensional coefficient of the second spherical harmonic,} \\
f &= 1/298.247, && \text{nondimensional value of the flattening.}
\end{aligned}
\tag{46}
$$

The values of GRS67 are smaller than the 1930 adopted international gravity formula (1930 IGF); it can be verified that (1930 IGF) – (1967 GRS) varies from 0.01715 gal at the equator to 0.00358 gal at the pole; this variation, however, is not linear. A slightly revised set of data was adopted in December 1979 and is known as GRS 1980. By using the Bruns formula [Eq. (38)], we can rewrite Eq. (36) as

$$\gamma(b + h;\, \theta_e) = \gamma(b;\theta_e)\left\{1 - h\left[\frac{1}{R_1} + \frac{1}{R_2} + \frac{2\omega^2}{\gamma(b;\theta_e)}\right]\right\}.$$

From Eqs. (17) and (18) we can also establish that

$$M = \frac{1}{R_1} + \frac{1}{R_2} = \frac{2b}{a^2}(1 + 2f\sin^2\theta_e),$$

and will make use of the gravity formula to obtain finally

$$\gamma(b + h;\theta_e) - \gamma(b;\theta_e) = -\frac{2h}{a}\gamma\left(b;\frac{\pi}{2}\right)[1 + f + 3q + (-3f + \tfrac{15}{2}q)\cos^2\theta_e]. \tag{47}$$

Here we are considering h, f, and q to be of the same order of magnitude, so that the coefficient of h is limited to the first powers of f and q.

The above expansion provides the normal component of gravity at an elevation h along the normal to the ellipsoid once we know the gravity at the foot of the normal on the ellipsoid and the gravity at the equator of the ellipsoid.

1.06 POTENTIAL OF THE ELLIPSOID EXPRESSED BY MEANS OF SPHERICAL HARMONICS

By making use of the series expansions for $\tan^{-1}(c/u)$ and for the function $Q(c/u)$ [see Eq. (31)], it is possible to represent the gravitational potential of the ellipsoid, which is given by Eq. (32), in terms of the ellipsoidal coordinates (u, θ_r, ϕ). After some algebraic manipulations, we obtain

$$U(b; u, \theta_r) = \frac{Gm_e}{u}\left\{1 + \sum_{n=1}^{\infty} \frac{(-1)^n}{(2n+1)}\left(\frac{c}{u}\right)^{2n} \times \left[1 - \frac{4n}{(2n+3)} \frac{qe_2}{Q(c/b)} P_2(\cos\theta_r)\right]\right\}, \tag{48}$$

where the parameters q and e_2 have already been introduced. Since

$$c/u = c/r + O(-3)$$

as a result of Eq. (23), it is possible to transform analytically the previous expression of the potential into the spherical harmonic representation of the same potential, which, from other considerations, is known to be of the form

$$U(a; r, \theta_g) = \frac{Gm_e}{r}\left[1 + \sum_{n=1}^{\infty} K_{2n}(a)\left(\frac{a}{r}\right)^{2n} P_{2n}(\cos\theta_g)\right]. \tag{49}$$

We can now use these two representations of the same potential to express the coefficients $K_{2n}(a)$ and the function $Q(c/b)$ in terms of physically significant quantities relating to the ellipsoid. For this purpose, let us consider a point of the rotational axis, for which we have $u = r$ and $\theta_r = \theta_g = 0$; since $P_n(1) = 1$ for any n, we can compare the two ensuing series in $1/r$ and obtain

$$K_{2n}(a) = \frac{(-1)^n e_1^{2n}}{2n+1}\left[1 - \frac{4n}{(2n+3)} \frac{qe_2}{Q(c/b)}\right]. \tag{50}$$

We know from elementary considerations that

$$K_2(a) = -\frac{(C - A)}{m_e a^2}, \tag{51}$$

where C is the moment of inertia of the body (ellipsoid) with respect to the rotational axis and A is the moment of inertia with respect to any equatorial barycentric axis. Comparing Eq. (51) with Eq. (50) evaluated for $n = 1$, we have

$$\frac{qe_2}{Q(c/b)} = \frac{5}{4}\left[1 - 3\left(\frac{C - A}{m_e c^2}\right)\right], \tag{52}$$

which provides a physical interpretation of the function $Q(c/b)$. Substituting Eq. (52) in Eq. (50), we can write

$$K_{2n}(a) = \frac{3(-e_1^2)^n}{(2n+1)(2n+3)}\left[1 - n + 5n\left(\frac{C - A}{m_e c^2}\right)\right]. \tag{53}$$

On the other hand, we can also use the known series expansion of $Q(c/b)$ to obtain from Eq. (52)

$$\frac{C - A}{m_e c^2} = \frac{1}{e_2^2}\left(\tfrac{1}{3}e_2^2 - q - \tfrac{6}{7}qe_2^2\right). \tag{54}$$

When this is substituted in Eq. (53) it gives rise to the expansions

$$\begin{aligned} K_2(a) &= -\left(\tfrac{2}{3}f - q - \tfrac{1}{3}f^2 + \tfrac{2}{7}fq\right), \\ K_4(a) &= \tfrac{4}{5}f^2 - \tfrac{12}{7}fq, \end{aligned} \tag{55}$$

up to second-order terms in f, q. Similar expressions for $K_{2n}(a)$ for $n > 2$ can be easily obtained.

1.07 STOKES FORMULA FOR GEOIDAL HEIGHTS

Let us consider the family of equipotential surfaces $V(x, y, z) = \text{const.}$ for the real earth (geopotential surfaces), of which the geoid corresponds to the constant value k_g; and the family of equipotential surfaces $U(x, y, z) = \text{const.}$ for the ellipsoid of revolution, from which a reference ellipsoid, of appropriate location, will be chosen, corresponding to the constant k_e.

Consider two elements of these families: $V = k_1$ and $U = k_2$, with $k_1 \neq k_2$. We can establish a correspondence between these two surfaces by saying that points P on $V = k_1$ and Q on $U = k_2$ are corresponding points when they are located on the same normal to the surface $U = k_2$. The normals to these surfaces will be denoted $\mathbf{n}_1$ and $\mathbf{n}_2$; the distance $\overrightarrow{QP}$, $N(Q)$ (see Fig. 1.4).

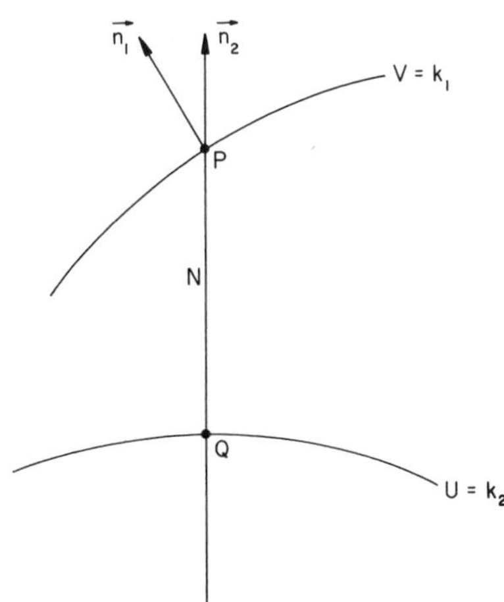

Fig. 1.4 Correspondence between points of a geopotential surface $V = \text{const.}$ and an ellipsoid of revolution $U = \text{const.}$

In the first-order approximation we have

$$U(P) = U(Q) + \left(\frac{\partial U}{\partial n_2}\right)_Q N(Q),$$

where the subscript Q indicates that the function has to be evaluated at Q. We shall call the difference $V(P) - U(P)$ at the same point in space the disturbing, or anomalous, potential, and denote it by $W(P)$. Note that $-\partial U/\partial n_2 = \gamma$ is the gravity of the ellipsoidal field and that $-\partial V/\partial n_1 = g$ is the gravity of the geoidal field. Using these concepts and the above notation, we can easily write

$$V(P) = U(P) + W(P) = U(Q) + W(P) - \gamma(Q)N(Q),$$

which can be simplified to

$$N(Q) = \frac{W(P) + k_2 - k_1}{\gamma(Q)}, \tag{56}$$

where k_1, k_2 are the two constants appearing in Fig. 1.4. This formula is due to Bruns and simply provides the normal distance between two equipotential surfaces in terms of the gravity field of one of them (the reference surface) and the supposedly known difference in potential functions and potential constants.

The gravity anomaly Δg at a given point P is defined as the difference between the values of the two gravity fields at the set of corresponding points (P, Q). Thus, the value of Δg at P, which we shall label $(\Delta g)(P)$, is equal to $g(P) - \gamma(Q)$.

We can now use the Bruns formula [Eq. (56)] to evaluate the gravity anomaly:

$$\begin{aligned}(\Delta g)(P) &= g(P) - \gamma(Q) = g(P) - \left[\gamma(P) - \left(\frac{\partial \gamma}{\partial n_2}\right)_Q N(Q)\right] \\ &= -\left(\frac{\partial V}{\partial n_1}\right)_P + \left(\frac{\partial U}{\partial n_2}\right)_P + \left(\frac{\partial \gamma}{\partial n_2}\right)_Q \frac{W(P) + k_2 - k_1}{\gamma(Q)} \\ &= -\left(\frac{\partial W}{\partial n_1}\right)_P + \frac{1}{\gamma(Q)}\left(\frac{\partial \gamma}{\partial n_2}\right)_Q [W(P) + k_2 - k_1].\end{aligned} \tag{57}$$

In obtaining this formula we have disregarded the angle δ between the two normals $\mathbf{n}_1$ and $\mathbf{n}_2$ since we are dealing with a first-order approximation; we have also used the concept of the disturbing potential $W = V - U$.

Equation (57) represents the fundamental formula of physical geodesy and is valid throughout space. It provides an approximate representation of the gravity anomaly existing at a set of corresponding points in terms of

(1) gravity and its normal derivative at a point on the reference surface and (2) the disturbing potential and its normal derivative evaluated at the corresponding point. In particular, the formula applies as a boundary condition on the geoid when we choose k_g and k_e as the constants corresponding to the geoid and to the reference ellipsoid.

Before proceeding any further, let us consider two simplifying assumptions which are physically plausible and which will reduce our future analytic developments to a manageable amount. We shall assume that there are no masses outside the geoid, so that the disturbing potential in that outer region is a harmonic function satisfying the Laplace equation for empty space. Next, in considering the reference ellipsoid we shall further simplify our computations, when dealing with small quantities such as W, N, and Δg, by neglecting the ellipsoidal flattening. This is tantamount to considering a spherical approximation to the ellipsoid assuming a mean radius r_s such that $r_s^3 = a^2 b$ and a gravity field γ_s such that $\gamma_s = Gm_e/r_s^2$; it follows that

$$\partial\gamma_s/\partial n_2 = \partial\gamma_s/\partial r = -2Gm_e/r_s^3$$

and

$$(\partial\gamma_s/\partial n_2)/\gamma_s = -2/r_s.$$

This is not, however, to be construed as if we had taken a sphere as the reference surface—Δg is always meant to be the difference between $g(P)$ and the ellipsoidal gravity $\gamma(Q)$; it is rather that we are considering a spherical approximation for the computations pertaining to the reference ellipsoid. With this type of approximation, the fundamental relation of physical geodesy becomes

$$(\Delta g)(P) + \left(\frac{\partial W}{\partial r}\right)_P + \frac{2}{r}[W(P) + k_2 - k_1] = 0. \tag{58}$$

The disturbing potential at $r = r_s$: $W(r_s, \theta, \phi)$, where θ is the colatitude and ϕ the longitude, can be expanded into spherical harmonics:

$$W(r_s; \theta, \phi) = \sum_{n=0}^{\infty} W_n(\theta, \phi).$$

In this connection, let us recall that the nth harmonic $f_n(\theta, \phi)$ in the spherical harmonic development of a given function $f(\theta, \phi)$ is given by

$$f_n(\theta, \phi) = \frac{2n+1}{4\pi} \iint_\sigma f(\theta', \phi') P_n(\cos\psi)\, d\sigma. \tag{59}$$

Here σ denotes the surface of the unit sphere and ψ is the angular distance between the fixed point P at which the function is evaluated and the variable integration point P'; see Fig. 1.5. We also have

$$\cos\psi = \cos\theta\cos\theta' + \sin\theta\sin\theta'\cos(\phi' - \phi), \tag{60}$$

and the surface element can be expressed as

$$d\sigma = \sin\theta'\, d\theta'\, d\phi' = \sin\psi\, d\psi\, d\alpha, \tag{61}$$

where α is the azimuthal angle.

For $r > r_s$, the disturbing potential can be expanded into spherical harmonics according to

$$W(r, \theta, \phi) = \sum_{n=0}^{\infty} \left(\frac{r_s}{r}\right)^{n+1} W_n(\theta, \phi).$$

This gives rise to

$$\frac{\partial W}{\partial r} = -\sum_{n=0}^{\infty} \frac{(n+1)}{r} \left(\frac{r_s}{r}\right)^{n+1} W_n(\theta, \phi),$$

and if we assume it to be convergent at $r = r_s$, we get

$$\frac{\partial W}{\partial r}(r_s, \theta, \phi) = -\frac{1}{r_s} \sum_{n=0}^{\infty} (n+1) W_n(\theta, \phi).$$

We can now write Eq. (58), with P taken on the geoid, $k_2 = k_e$, and $k_1 = k_g$, as

$$(\Delta g)(P) = \frac{1}{r_s} \sum_{n=0}^{\infty} (n-1) W_n(\theta, \phi) + \frac{2}{r_s}(k_g - k_e).$$

This relation provides the connection between the spherical harmonic developments for Δg and $W(r_s, \theta, \phi)$. Specifically, we realize that

$$\begin{aligned}
(\Delta g)_0 &= -\frac{1}{r_s} W_0 + \frac{2}{r_s}(k_g - k_e), \\
(\Delta g)_1 &= 0, \\
(\Delta g)_n &= \frac{n-1}{r_s} W_n(\theta, \phi), \qquad n \geqslant 2.
\end{aligned} \tag{62}$$

Thus, by measuring Δg and evaluating its spherical harmonic development, we have a way of representing the disturbing potential $W(r_s, \theta, \phi)$.

Recalling Eqs. (59) and (62), we can write

$$W(r_s, \theta, \phi) = W_0 + \frac{r_s}{4\pi} \iint_\sigma \left[\sum_{n=2}^{\infty} \frac{2n+1}{n-1} P_n(\cos\psi)\right] (\Delta g)\, d\sigma, \tag{63}$$

Library
I.U.P.
Indiana, Pa.

551.1 L297d
C.1

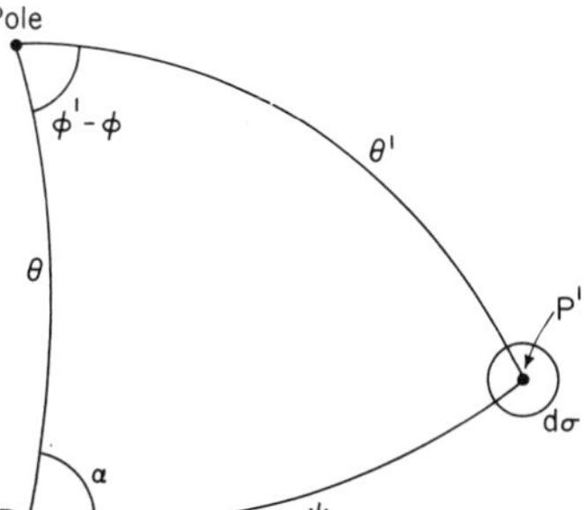

Fig. 1.5 Evaluation of the Stokes integral formula for the point P by averaging the gravity anomaly along a parallel of constant spherical radius.

where the first-degree harmonic W_1 is missing because we are assuming that the center of the reference ellipsoid coincides with Earth's center of gravity.

We introduce the Stokes function

$$S(\cos\psi) = \sum_{n=2}^{\infty} \frac{2n+1}{n-1} P_n(\cos\psi), \tag{64}$$

and use the Bruns formula to write

$$N(Q) = -\frac{(k_g - k_e)}{\gamma_s} + \frac{1}{\gamma_s}\left[W_0 + \frac{r_s}{4\pi}\iint_\sigma (\Delta g) S(\cos\psi)\, d\sigma\right].$$

Since

$$W_0 = \frac{G(m_g - m_e)}{r_s},$$

we obtain

$$N(Q) = N_0 + N_1(Q). \tag{65}$$

Here

$$N_0 = \frac{G(m_g - m_e)}{r_s \gamma_s} - \frac{(k_g - k_e)}{\gamma_s} \tag{66}$$

is the constant term, where m_g and m_e are the masses of the geoid and reference ellipsoid, respectively; the k are the equipotential constants; r_s and γ_s have the following values:

$$r_s = 6{,}371 \quad \text{km}, \qquad \gamma_s = 979.8 \quad \text{gal}; \tag{67}$$

and

$$N_1 = \frac{r_s}{4\pi\gamma_s}\iint_\sigma (\Delta g) S(\cos\psi)\, d\sigma \tag{68}$$

represents the integral contribution. Equations (65)–(68) constitute the Stokes formula for the geoidal heights.

In Chapter IV, we shall evaluate the Stokes and other related functions in finite terms by making use of the Poisson sum of a Legendre series.

At present it will suffice to state that

$$S(\cos\psi) \equiv \operatorname{cosec}(\psi/2) - 6\sin(\psi/2) + 1 - 5\cos\psi - 3(\cos\psi)\ln[\sin(\psi/2) + \sin^2(\psi/2)]. \tag{69}$$

Computationally, it is advantageous to treat the two integrals appearing in the Stokes formula in the following order:

$$N_1 = \frac{r_s}{2\gamma_s}\int_{\psi=0}^{\pi}\left[\frac{1}{2\pi}\int_{\alpha=0}^{2\pi}(\Delta g)\,d\alpha\right]S(\cos\psi)\sin\psi\,d\psi, \tag{70}$$

where we emphasize the average value of Δg along a parallel of constant spherical radius ψ; see Fig. 1.5. We can show that the function $S(\cos\psi)\sin\psi$ remains finite as ψ approaches zero. The only term that requires a special handling is $(\sin\psi)\ln[\sin(\psi/2)]$, and we can use the l'Hospital formula for indeterminate forms to show that it approaches zero as $\sin(\psi/2)$.

1.08 DEFLECTION OF THE VERTICAL AND THE FORMULA OF VENING-MEINESZ

Corresponding to the Stokes formula for the geoidal undulation, there exists another formula that provides the deflection of the vertical in terms of an integral of the gravity anomaly. By deflection of the vertical we mean the angle δ between the normal $\mathbf{n}_1$ to the geoid and the normal $\mathbf{n}_2$ to the reference ellipsoid. This formula was obtained by Vening-Meinesz in 1928.

Consider a plane, normal to the reference ellipsoid at a point Q, having an arbitrary azimuth α_e from the north. Let ϵ be the angle between the normal to the ellipsoid and the projection $\mathbf{n}_{1,\alpha}$ of the geoidal normal onto this plane. From Fig. 1.6 we realize that in first-order approximation we have

$$dN = |\epsilon|\,ds_1, \tag{71}$$

where ds_1 is the element of arc along the geoid; the sign above can be chosen according to an appropriate convention.

Before proceeding any further along these lines, let us first establish how the angle ϵ, which is the component of the vertical deflection upon the normal plane of azimuth α, depends on the component ξ of the vertical deflection in the north–south direction and on the component η of the vertical deflection in the east–west direction. As a by-product of this analysis we shall ascertain the azimuthal variation of a target point when we compare measurements

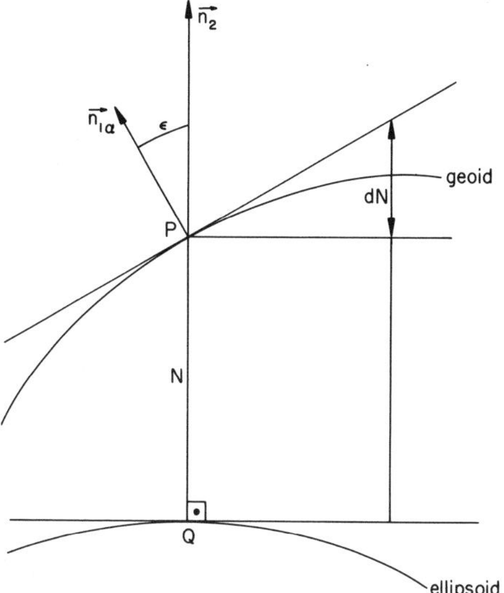

Fig. 1.6 Geoidal height N upon the reference ellipsoid and component ϵ of the vertical deflection upon a normal plane of given azimuth.

performed on the geoid with measurements pertaining to the ellipsoid. For this purpose, we consider the unit sphere at the observation station P on the geoid and the tangent plane to the geoid at P which intersects the sphere according to the horizon circle of P. A view from the top in the direction of the geoidal normal provides the projection of all the vectors involved onto the tangent plane; see Fig. 1.7.

We are considering four vectors: (1) $\overrightarrow{PP_N}$ to the North Pole P_N, elevated by the angle $(\pi/2) - \theta_a$ above the horizon; (2) $\overrightarrow{PZ_a}$, the geoidal normal; (3) $\overrightarrow{PZ_e}$, the ellipsoidal normal; and (4) $\overrightarrow{Z_aT}$ and $\overrightarrow{Z_eT}$, the directions to the target point. We are supposing that the target T is practically on the horizon, so that both zenithal distances $z_1 = \overrightarrow{Z_aT}$ and $z_2 = \overrightarrow{Z_eT}$ are nearly 90°.

The vertical deflection $\overrightarrow{Z_aZ_e}$ will have the component $\xi = \overrightarrow{Z_aF}$ in the north–south direction and $\eta = \overrightarrow{Z_eF}$ in the east–west direction. $\delta^2 = \xi^2 + \eta^2 = (\overrightarrow{Z_aZ_e})^2$ is the square of the vertical deflection. $\overrightarrow{Z_eG} = \epsilon$ is the component of the vertical deflection upon the normal plane having the ellipsoidal azimuth α_e. We plan to obtain expressions for ξ, η, ϵ, and the azimuthal variation $\alpha_a - \alpha_e$ by means of spherical trigonometry and the small angles approximation.

From the right spherical triangle P_NFZ_e, we have

$$\frac{\sin(Z_eF)}{\sin(\Delta\phi)} = \frac{\eta}{\Delta\phi} = \frac{\sin\theta_e}{\sin 90^\circ},$$

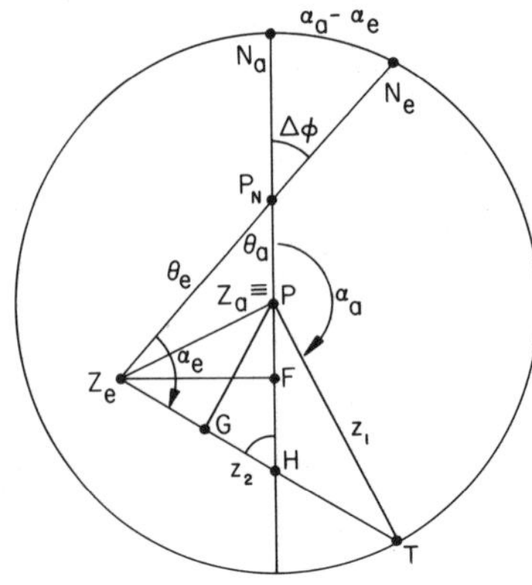

Fig. 1.7 Azimuth variation for a fixed target point T encountered when one reduces measurements made on the geoid to measurements on the reference ellipsoid.

or

$$\eta = (\Delta\phi)\sin\theta_e. \tag{72}$$

From the same triangle we can also write

$$\cos\theta_e = \cos(\theta_a - \xi)\cos\eta,$$

and since $\cos\eta \sim 1$, we get $\theta_e = \theta_a - \xi$, or

$$\xi = \theta_a - \theta_e. \tag{73}$$

From the spherical triangle $P_N N_e N_a$ we have

$$\frac{\sin(\alpha_a - \alpha_e)}{\sin(\Delta\phi)} = \frac{\alpha_a - \alpha_e}{\Delta\phi} = \frac{\cos\theta_e}{\sin 90^\circ},$$

or using Eq. (72),

$$\alpha_a - \alpha_e = (\Delta\phi)\cos\theta_e = \eta\cot\theta_e. \tag{74}$$

This is known as the Laplace formula for the azimuthal reduction. Finally we have

$$\epsilon = \overrightarrow{Z_e G} = \eta\sin(\hat{H}) - \xi\cos(\hat{H}),$$

where the angle at H can be reckoned from the flat triangle approximation to be

$$\hat{H} = \pi - (\Delta\phi + \alpha_e) \cong \pi - \alpha_e.$$

This gives

$$\epsilon = \eta\sin\alpha_e + \xi\cos\alpha_e. \tag{75}$$

Let us emphasize that in Fig. 1.7, the distance $\delta = \overrightarrow{Z_a Z_e}$ is greatly exaggerated for sake of clarity; both $\Delta\phi$ and $\Delta\theta$ are small angles.

Let us now revert to Eq. (71). Along the north–south direction ($\alpha_a = 0$, $\Delta\phi = 0$), we have $\epsilon = \xi$ and $ds_1 = r_s\, d\theta_a$. In the east–west direction, we have $\epsilon = \eta$ and $ds_1 = r_s \sin\theta_a\, d\phi_a$. This is so because on a sphere of radius r_s, the linear element is given by $(ds_1)^2 = r_s^2(d\theta_a)^2 + r_s^2 \sin^2\theta_a (d\phi_a)^2$, where θ_a is the colatitude. We can then write

$$|\xi| = \frac{dN}{ds_1} = \frac{1}{r_s}\frac{\partial N}{\partial\theta},$$

$$|\eta| = \frac{dN}{ds_1} = \frac{1}{r_s \sin\theta}\frac{\partial N}{\partial\phi},$$

and N is given by the Stokes formula. We rewrite the variable portion N_1 of the Stokes formula as

$$N_1(\theta, \phi) = \frac{r_s}{4\pi\gamma_s}\int_0^{2\pi}\int_0^{\pi}(\Delta g)(\theta', \phi')S(\cos\psi)\sin\theta'\, d\theta'\, d\phi',$$

where the primed quantities are the integration variables; see Fig. 1.5. The variables θ and ϕ appear in the integrand through $\cos\psi$, which is given by Eq. (60). We need then to evaluate

$$\frac{\partial S}{\partial\theta} = \frac{dS}{d\psi}\frac{\partial\psi}{\partial\theta}, \qquad \frac{\partial S}{\partial\phi} = \frac{dS}{d\psi}\frac{\partial\psi}{\partial\phi}.$$

Differentiation of Eq. (60) yields

$$-\sin\psi\,\frac{\partial\psi}{\partial\theta} = -\sin\theta\cos\theta' + \cos\theta\sin\theta'\cos(\phi' - \phi),$$

$$-\sin\psi\,\frac{\partial\psi}{\partial\phi} = \sin\theta\sin\theta'\sin(\phi' - \phi).$$

We must also consider the spherical triangle in Fig. 1.5 and apply two formulas concerning the azimuth α:

$$\sin\psi\cos\alpha = \sin\theta\cos\theta' - \cos\theta\sin\theta'\cos(\phi' - \phi),$$
$$\sin\psi\sin\alpha = \sin\theta'\sin(\phi' - \phi).$$

Comparison of the last two sets of formulas gives rise to

$$\frac{\partial\psi}{\partial\theta} = \cos\alpha, \qquad \frac{\partial\psi}{\partial\phi} = -\sin\theta\sin\alpha.$$

Ultimately we obtain

$$|\xi| = \frac{1}{4\pi\gamma_s}\iint_\sigma (\Delta g)\frac{dS}{d\psi}\cos\alpha\, d\sigma,$$
$$|\eta| = \frac{1}{4\pi\gamma_s}\iint_\sigma (\Delta g)\frac{dS}{d\psi}\sin\alpha\, d\sigma, \tag{76}$$

where the element of area is given by Eq. (61). Differentiation of the Stokes function gives

$$\begin{aligned} dS/d\psi = {} & -\tfrac{1}{2}[\cos(\psi/2)/\sin^2(\psi/2)] - 3\cos(\psi/2) \\ & + 5\sin\psi + 3(\sin\psi)\ln[\sin(\psi/2) + \sin^2(\psi/2)] \\ & - 3(\cos\psi)[\tfrac{1}{2}\cos(\psi/2) + \sin(\psi/2)\cos(\psi/2)]/[\sin(\psi/2) + \sin^2(\psi/2)]. \end{aligned}$$

It is more convenient to simplify the last term by simple trigonometric manipulations to yield

$$-3\cos(\psi/2) + 3\sin\psi - 3[1 - \sin(\psi/2)]/\sin\psi.$$

Equation (76) represents the Vening-Meinesz formula. Let us write this formula as

$$\begin{bmatrix}|\xi|\\|\eta|\end{bmatrix} = \frac{1}{2\gamma_s}\int_0^\pi\left[\frac{1}{2\pi}\int_0^{2\pi}(\Delta g)\begin{pmatrix}\cos\alpha\\ \sin\alpha\end{pmatrix}d\alpha\right]\frac{dS}{d\psi}\sin\psi\, d\psi.$$

In the neighborhood of the point P the integrand appears indeterminate. In fact, the function $(dS/d\psi)\sin\psi$ becomes infinite like $1/\sin(\psi/2)$ as $\psi \to 0$; on the other hand, for small values of ψ, we have

$$\int_0^{2\pi}(\Delta g)\begin{pmatrix}\cos\alpha\\ \sin\alpha\end{pmatrix}d\alpha \cong (\Delta g)(P)\int_0^{2\pi}\begin{pmatrix}\cos\alpha\\ \sin\alpha\end{pmatrix}d\alpha = 0.$$

To resolve this indeterminate expression we shall, in the next section, replace Δg by a series expansion and replace the spherical cap around P by its tangent plane.

1.09 COMPUTATIONAL TECHNIQUES

Let us investigate certain computational procedures which are appropriate for the evaluation of the two previous formulas over the whole earth. We shall elaborate on (1) the computation of the integrals in the neighborhood of the point P and (2) the contribution that distant zones have on the overall value of the integrals.

In the immediate neighborhood of the position P, where the functions have to be evaluated, we can replace Δg by a series expansion and also the spherical cap around P by its tangent plane. For the sake of simplicity, we shall denote the gravity anomaly Δg by G and consider its Taylor series expansion about P:

$$\begin{aligned} G(P') = {} & G(P) + xG_x(P) + yG_y(P) \\ & + \tfrac{1}{2}[x^2 G_{xx}(P) + 2xyG_{xy}(P) + y^2 G_{yy}(P)] + \cdots, \end{aligned}$$

where $x = s\cos\alpha$ and $y = s\sin\alpha$, α being the azimuth; the subscripts here denote partial derivatives. The Stokes function expansion is

$$S(\cos\psi) = (2/\psi) + \cdots = 2(r_s/s) + \cdots$$

since in this approximation $s = r_s\psi$; thus

$$dS/d\psi = -(2/\psi^2) + \cdots = -2(r_s/s)^2 + \cdots$$

and

$$\begin{aligned} d\sigma &= \sin\psi \, d\psi \, d\alpha \cong \psi \, d\psi \, d\alpha \\ &= (s/r_s)(ds/r_s) \, d\alpha = (s/r_s^2) \, ds \, d\alpha. \end{aligned}$$

Hence

$$\begin{aligned} N_1(P') = {} & \frac{1}{2\pi\gamma_s} \int_0^{2\pi} \int_0^{s_0} \{G(P) + s[G_x(P)\cos\alpha + G_y(P)\sin\alpha] \\ & + \tfrac{1}{2}s^2[G_{xx}(P)\cos^2\alpha + 2G_{xy}(P)\sin\alpha\cos\alpha \\ & + G_{yy}(P)\sin^2\alpha]\} \, ds \, d\alpha + \cdots \\ = {} & \frac{s_0 G(P)}{\gamma_s} + \frac{1}{12\gamma_s} s_0^3[G_{xx}(P) + G_{yy}(P)] + \cdots \end{aligned} \tag{77}$$

because the integrals of $\sin\alpha$, $\cos\alpha$, and $\sin\alpha\cos\alpha$ in the interval $(0, 2\pi)$ vanish; whereas the integrals of $\sin^2\alpha$ and $\cos^2\alpha$ in the same interval amount to π. Similarly, we obtain

$$\begin{aligned} \begin{bmatrix} \xi(P') \\ \eta(P') \end{bmatrix} = {} & -\frac{1}{2\pi\gamma_s} \int_0^{2\pi} \int_0^{s_0} \left[\frac{G(P)}{s} + G_x(P)\cos\alpha \right. \\ & \left. + G_y(P)\sin\alpha + \tfrac{1}{2}s(\cdots) \right] \begin{pmatrix} \cos\alpha \\ \sin\alpha \end{pmatrix} ds \, d\alpha \\ = {} & -\frac{1}{2\gamma_s} \begin{bmatrix} G_x(P) \\ G_y(P) \end{bmatrix} s_0 + s_0^2(\cdots) + \cdots. \end{aligned} \tag{78}$$

The values of $G(P)$, $G_x(P)$, $G_y(P)$ can be obtained from maps of Δg and the gradient of Δg in that particular region of interest.

The technique just examined addresses the problem of computation around the point P where N_1 and (ξ, η) have to be evaluated.

Another question of practical importance is to account for the influence that distant zones (from P) have in the evaluation of the integral formula. For the sake of brevity we shall discuss only the Stokes formula.

We can account for the contribution to the integral beyond a spherical cap of semiaperture ψ_0 centered at P in the following fashion.

We define a new Stokes function:

$$S^*(\cos\psi) = \begin{cases} 0 & \text{for} \quad 0 \leqslant \psi < \psi_0 \\ S(\cos\psi) & \text{for} \quad \psi_0 \leqslant \psi \leqslant \pi. \end{cases}$$

Then

$$\delta N_1 = \frac{r_s}{4\pi\gamma_s} \int_0^{2\pi} \int_0^{\pi} (\Delta g) S^*(\cos\psi) \sin\psi \, d\psi \, d\alpha.$$

Let us consider the spherical harmonic representation of S^*:

$$S^*(\cos\psi) = \sum_{n=0}^{\infty} S_n P_n(\cos\psi),$$

where the coefficients of the various harmonics are, according to Eq. (59),

$$\begin{aligned} S_n &= \frac{2n+1}{4\pi} \int_0^{2\pi} \int_0^{\pi} S^*(\cos\psi) P_n(\cos\psi) \sin\psi \, d\psi \, d\alpha \\ &= \frac{2n+1}{2} \int_0^{\pi} S^*(\cos\psi) P_n(\cos\psi) \sin\psi \, d\psi \\ &= \frac{2n+1}{2} \int_{\psi_0}^{\pi} S(\cos\psi) P_n(\cos\psi) \sin\psi \, d\psi. \end{aligned}$$

It follows then that

$$\begin{aligned} \delta N_1 &= \frac{r_s}{4\pi\gamma_s} \sum_{n=0}^{\infty} S_n \int_0^{2\pi} \int_0^{\pi} (\Delta g) P_n(\cos\psi) \sin\psi \, d\psi \, d\alpha \\ &= \frac{r_s}{4\pi\gamma_s} \sum_{n=0}^{\infty} S_n \frac{4\pi}{(2n+1)} (\Delta g)_n \\ &= \frac{r_s}{\gamma_s} \sum_{n=0}^{\infty} \frac{S_n}{(2n+1)} (\Delta g)_n \end{aligned} \tag{79}$$

since the last integral is proportional to the coefficients of the spherical harmonic representation of Δg. The expression provides the error in the geoidal height at a given point $P(\theta, \phi)$, which is incurred should one neglect the gravity anomaly beyond a circular cap of radius ψ_0 centered at P. It requires the knowledge of the spherical harmonic representation of Δg and the functions $S_n(\cos\psi_0)$. Let us briefly discuss the latter quantities. If we introduce new variables

$$z = \sin(\psi/2), \qquad t = \sin(\psi_0/2),$$

we have

$$S_n(t) = -2(2n+1)\int_1^t S(1-2z^2)P_n(1-2z^2)z\,dz. \tag{80}$$

The Stokes function $S(1-2z^2)$ is easily seen to be

$$(1/z) + (6z^2 - 3)\ln[z(1+z)] - 4 - 6z + 10z^2.$$

Since $P_n(1-2z^2)$ is a polynomial in z^2, in order to evaluate $S_n(t)$ we must integrate functions such as $z^n \ln z$ and $z^n \ln(1+z)$. This can be done in closed form by elementary methods. In fact,

$$\int z^n \ln z\,dz = \frac{z^{n+1}}{n+1}\ln z - \frac{z^{n+1}}{(n+1)^2}$$

and

$$\int z^n \ln(1+z)\,dz = \frac{z^{n+1}}{n+1}\ln(1+z) - \frac{1}{(n+1)}\int \frac{z^{n+1}}{z+1}\,dz.$$

We can integrate the last expression by using the algebraic identity

$$\begin{aligned}\frac{z^n}{z+1} &= \frac{z^n - (-1)^n}{z+1} + \frac{(-1)^n}{z+1}\\ &= z^{n-1} - z^{n-2} + \cdots + (-1)^{n-2}z + (-1)^{n-1} + \frac{(-1)^n}{z+1}.\end{aligned}$$

This simple discussion shows that $S_n(t)$ must be of the form

$$S_n(t) = p(t) + q(t)\ln t + r(t)\ln(1+t),$$

where $p(t)$, $q(t)$, and $r(t)$ are polynomials in t. Since $S_n(1) = 0$ because of Eq. (80), we have

$$S_n(1) = 0 = p(1) + r(1)\ln 2;$$

we conclude that the above expression cannot vanish for the reason that $\ln 2$ (which is an irrational quantity) could become equal to $-p(1)/r(1)$

(which is a rational quantity). The expression can only vanish identically in the sense that $p(1)$ and $r(1)$ must be zero separately. This means that the sum of all the coefficients appearing within each of these polynomials must add up to zero; the result can be used for verification purposes. For further details on this procedure, one should consult Molodenskii *et al.* (1962).

In the Stokes formula we do not know the values of m_g and k_g well enough to reach an accurate determination of the term N_0. The evaluation of the integral term N_1 according to the methods exposed in this section will allow the determination of a surface Σ^* parallel to the geoid Σ at a distance N_0 from it. The scale factor between these two surfaces can be ascertained by performing a single distance measurement on the geoid.

For this purpose, consider an arc ds_1 on the geoid and the corresponding arc ds_2 on the reference ellipsoid. Using the spherical approximation already discussed, we have from Fig. 1.6

$$(ds_1)\cos\epsilon/(r_s + N) = ds_2/r_s.$$

Assuming that $\cos\epsilon \sim 1$, by integration we reach the result

$$s_1 = s_2 + \frac{1}{r_s}\int_{Q_1}^{Q_2}(N_0 + N_1)\,ds_2,$$

or

$$N_0 = \frac{r_s}{s_2}(s_1 - s_2) - \frac{1}{s_2}\int_{Q_1}^{Q_2} N_1\,ds_2. \tag{81}$$

The quantity N_1 is given by the Stokes integral. We must perform the length measurement s_1 on the geoid. In actuality this measurement will be performed over the topography of the real earth and must therefore somehow be reduced to the geoid; however, we shall not deal with this problem. We must also determine the arc s_2 on the reference ellipsoid between the points Q_1 and Q_2. By means of astronomical observations we can determine the direction of the plumb line (i.e., the normal $\mathbf{n}_1$ to the geoid) or equivalently the astronomical colatitude θ_a and astronomical longitude ϕ_a. Next, the use of the Vening-Meinesz formula allows us to ascertain the components (ξ, η) of the vertical deflection in the north–south and east–west directions; see Eq. (76). Finally, we use Eqs. (72) and (73), that is,

$$\xi = \theta_a - \theta_e \qquad \text{and} \qquad \eta = (\phi_a - \phi_e)\sin\theta_e,$$

to determine the ellipsoidal colatitude θ_e and longitude ϕ_e. We can then calculate the arc s_2 on the ellipsoid and eventually N_0 from Eq. (81). In practice, many distances and angles are measured, and a suitable average will provide the value of N_0. In this connection, the deformation of the azimuthal coordinate in passing from geoid to ellipsoid [the Laplace formula, given by Eq. (74)] must be used.

We can now use this independent evaluation of N_0 to reach a more accurate value for the mass m_g of the geoid and for its constant potential k_g. For this purpose, we shall choose an arbitrary reference ellipsoid of known mass m_e and known constant potential k_e. We then compute the gravity anomaly Δg with respect to this ellipsoid and numerically evaluate

$$(\Delta g)_0 = \frac{1}{4\pi} \iint_\sigma \Delta g \, d\sigma,$$

which is the zero-order harmonic in the spherical harmonic expansion of Δg. We now consider the set of formulas given by Eqs. (62) and (66):

$$N_0 = \frac{G(m_g - m_e)}{r_s \gamma_s} - \frac{k_g - k_e}{\gamma_s},$$

$$(\Delta g)_0 = -\frac{1}{r_s^2} G(m_g - m_e) + \frac{2}{r_s}(k_g - k_e).$$

We know the following quantities: N_0, $(\Delta g)_0$, m_e, k_e, r_s, γ_s, G; we can solve linearly for m_g and k_g, obtaining

$$\begin{aligned} k_g &= k_e + N_0 \gamma_s + r_s (\Delta g)_0, \\ m_g &= m_e + \frac{r_s}{G} [2N_0 \gamma_s + r_s (\Delta g)_0]. \end{aligned} \tag{82}$$

1.10 DETERMINATION OF THE GEOID BY ASTROGEODETIC METHODS: HELMERT FORMULA AND SHIFT IN GEODETIC DATUM

As a result of Eqs. (71) and (75),

$$dN = |\epsilon| \, ds_1 = (\xi \cos \alpha_e + \eta \sin \alpha_e) \, ds_1.$$

We integrate and get

$$N(P_2) = N(P_1) + \int_{P_1}^{P_2} (\xi \cos \alpha_e + \eta \sin \alpha_e) \, ds_1. \tag{83}$$

This formula is known as Helmert's formula and expresses the geoidal undulation as a line integral of the vertical deflection along a profile. The integral is independent of the path followed between P_1 and P_2 because N is a function of position alone. The path followed need not be a geodesic of the ellipsoid and, in practice, either a north–south or east–west profile is chosen. The integral has to be evaluated by numerical integration from maps of ξ and η pertaining to a certain region where numerical interpolation between data is feasible.

The components of the vertical deflection can be obtained either from the Vening-Meinesz formula using the gravimetric method or from the astrogeodetic method by means of the differences between astronomical and ellipsoidal coordinates as expressed by Eqs. (72) and (73). The astronomical coordinates θ_{a}, ϕ_{a} can be determined by direct observations. The ellipsoidal coordinates are obtainable by considering a triangulation net and reducing the measured angles and distances from the geoid to the ellipsoid; then one determines on the ellipsoid the nodes of the net and computes their ellipsoidal coordinates. Comparison with the measured astronomical coordinates according to the equations mentioned provides the input ξ, η for the line integral.

It is essential that a specific ellipsoid of semimajor axis a and ellipticity f be assigned at one node of the net; its relative position with respect to the geoid can be ascertained by the values of N, ξ, η at that node with the proviso that the axis of revolution of the reference ellipsoid be parallel to Earth's rotational axis. This set of five parameters $(a, f; N; \xi, \eta)$ or any equivalent set is called a geodetic datum.

Both Helmert's and Stokes's formulas deal with the gravity anomaly vector $\Delta\mathbf{g}$: in one case we use its components (ξ, η); in the other case, its magnitude Δg. Both formulas contain these elements linearly. The essential difference is as follows.

The Helmert formula makes use of any profile on land, is limited to a bounded area, and is applicable to a reference ellipsoid whose position with respect to Earth's center of gravity is unknown.

In the Stokes formula, the integral is extended over the whole Earth—it includes sea measurements; the ellipsoid of reference is known with respect to the center of gravity of Earth, but its scale must be determined by the astrogeodetic method. Thus, the two methods are interrelated; ultimately, satellite geodesy can play a useful role in supplementing these results.

One fundamental problem that must be solved in the application of the astrogeodetic method is the shifting of the local geodetic datum from one location to another neighboring position. We can solve this problem by means of an infinitesimal coordinate transformation.

We consider a reference ellipsoid having its axis of revolution parallel to Earth's rotational axis and whose center is located at (x_0, y_0, z_0) with respect to Earth's center. The z-coordinate of a point Q on the ellipsoid (see Fig. 1.8) is $R_2 \cos\theta_{\mathrm{e}} + \overrightarrow{OT}$, where according to Section 1.2, $OT = -zc^2/b^2$; combining the two results, we have $z = (b^2/a^2)R_2 \cos\theta_{\mathrm{e}}$.

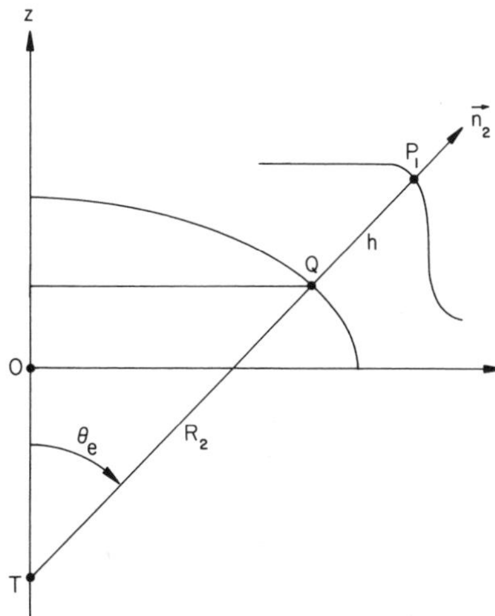

Fig. 1.8 Geometric height h of a point on Earth from reference ellipsoid along the ellipsoidal normal.

The cartesian coordinates of any point P_1 on Earth, of geometric height $h = QP_1$, measured along $\mathbf{n}_2$, can be written as

$$\begin{aligned} x &= x_0 + (R_2 + h)\sin\theta_e\cos\phi_e, \\ y &= y_0 + (R_2 + h)\sin\theta_e\sin\phi_e, \\ z &= z_0 + \left(\frac{b^2}{a^2}R_2 + h\right)\cos\theta_e. \end{aligned} \tag{84}$$

We plan to investigate how these equations vary when we allow a variation of a, f (dimensions of the ellipsoid) and of h, θ_e, ϕ_e (its location with respect to the geoid). Let us first consider some series expansions of the quantities in question. From Eq. (18),

$$\begin{aligned} R_2 &= \frac{a^2}{b}(1 + e_2^2\sin^2\theta_e)^{-1/2} = a\frac{a}{b}(1 - \tfrac{1}{2}e_2^2\sin^2\theta_e + \cdots) \\ &= a(1+f)(1 - f\sin^2\theta_e + \cdots) = a(1 + f\cos^2\theta_e) + \cdots, \end{aligned}$$

to the first power of f, since $b/a = 1 - f$ and $e_2^2 = 2f + \cdots$. Similarly,

$$\frac{b^2}{a^2}R_2 = (1-2f)a(1 + f\cos^2\theta_e) + \cdots = a(1 - 2f + f\cos^2\theta_e) + \cdots.$$

To the same order of approximation, the cartesian coordinates of P_1 can be written as

$$\begin{aligned} x &= x_0 + (a + af\cos^2\theta_e + h)\sin\theta_e\cos\phi_e, \\ y &= y_0 + (a + af\cos^2\theta_e + h)\sin\theta_e\sin\phi_e, \\ z &= z_0 + (a - 2af + af\cos^2\theta_e + h)\cos\theta_e. \end{aligned} \tag{85}$$

Let us now consider Taylor series expansions of these formulas up to first-order terms:

$$\delta x = \delta x_0 + \frac{\partial x}{\partial a}\,\delta a + \frac{\partial x}{\partial f}\,\delta f + \frac{\partial x}{\partial \theta_e}\,\delta\theta_e + \frac{\partial x}{\partial \phi_e}\,\delta\phi_e + \frac{\partial x}{\partial h}\,\delta h. \tag{86}$$

Two similar relations hold for the other two coordinates. Table 1.1 represents the 3×5 matrix of the differential coefficients; in evaluating them, we have considered the terms f, $f \cos \theta_e$, and h negligible when they are additive to a.

Table 1.1

Increments of the Cartesian Coordinates of a Point on Earth Due to a Variation in the Dimensions of the Reference Ellipsoid and Its Location with Respect to the Geoid

	δa	$a\,\delta f$	$a\,\delta\theta$	$a\,\delta\phi$	δh
δx	$\sin\theta\cos\phi$	$\cos^2\theta\sin\theta\cos\phi$	$\cos\theta\cos\phi$	$-\sin\theta\sin\phi$	$\sin\theta\cos\phi$
δy	$\sin\theta\sin\phi$	$\cos^2\theta\sin\theta\sin\phi$	$\cos\theta\sin\phi$	$\sin\theta\cos\phi$	$\sin\theta\sin\phi$
δz	$\cos\theta$	$(\cos^2\theta - 2)\cos\theta$	$-\sin\theta$	0	$\cos\theta$

These formulas express the variation $(\delta x, \delta y, \delta z)$ of the cartesian coordinates of a point due to a change in (1) the translational position $(\delta x_0, \delta y_0, \delta z_0)$ of the ellipsoid; (2) its dimension $(\delta a, \delta f)$, and (3) the location $(\delta\theta, \delta\phi, \delta h)$ at which it is connected to the geoid.

The fundamental problem of shifting the geodetic datum can be formulated as follows.

We change the dimensions of the reference ellipsoid ($\delta a \neq 0$, $\delta f \neq 0$) retaining, however, its rotational axis parallel to Earth's axis of rotation; and we attach the reference ellipsoid to a different point P_1 on Earth ($\delta\theta_1 \neq 0$, $\delta\phi_1 \neq 0$, $\delta h_1 \neq 0$); we then seek to ascertain the coordinates $\delta\theta$, $\delta\phi$, δh with respect to this new reference ellipsoid for a fixed point P on Earth ($\delta x = \delta y = \delta z = 0$).

The execution of these operations is equivalent to a repeated solution of the linear system (86), whose coefficient matrix is furnished by Table 1.1. Three consecutive steps must be taken:

1. First, we solve Eq. (86) for $\delta x_0, \delta y_0, \delta z_0$ in terms of $\delta\theta_1, \delta\phi_1, \delta h_1$, δa, δf, assuming that $\delta x = \delta y = \delta z = 0$.
2. Next, we solve the same system for $\delta\theta$, $\delta\phi$, δh in terms of $\delta x_0, \delta y_0$, δz_0, δa, δf, assuming again that $\delta x = \delta y = \delta z = 0$.
3. Finally, we eliminate the $\delta x_0, \delta y_0, \delta z_0$ appearing in the previous step by using the results obtained in step (1).

Table 1.2

Variation of the Nondimensional Parameters Due to a Shift in Geodetic Datum. Note That $\Delta\phi = \phi - \phi_1$

	$(\delta a)/a$	δf	$\delta\xi_1$	$\delta\eta_1$	$(\delta N_1)/a$
$\delta\xi$	$\sin\theta\cos\theta_1 - \sin\theta_1\cos\theta\cos(\Delta\phi)$	$\cos^2\theta_1[\sin\theta\cos\theta_1 - \sin\theta_1\cos\theta\cos(\Delta\phi)] + 2\sin\theta(\cos\theta - \cos\theta_1)$	$\sin\theta\sin\theta_1 + \cos\theta_1\cos\theta\cos(\Delta\phi)$	$\cos\theta\sin(\Delta\phi)$	$\sin\theta\cos\theta_1 - \sin\theta_1\cos\theta\cos(\Delta\phi)$
$\delta\eta$	$\sin\theta_1\sin(\Delta\phi)$	$\cos^2\theta_1\sin\theta_1\sin(\Delta\phi)$	$-\cos\theta_1\sin(\Delta\phi)$	$\cos(\Delta\phi)$	$\sin\theta_1\sin(\Delta\phi)$
$\frac{\delta N}{a}$	$\sin\theta\sin\theta_1\cos(\Delta\phi) + \cos\theta\cos\theta_1 - 1$	$\cos^2\theta_1\sin\theta_1\sin\theta\cos(\Delta\phi) + \cos\theta\cos\theta_1(\cos^2\theta_1 - 2) + \cos^2\theta$	$\cos\theta\cos\theta_1 - \cos\theta_1\sin\theta\cos(\Delta\phi)$	$-\sin\theta\sin(\Delta\phi)$	$\cos\theta\cos\theta_1 + \sin\theta\sin\theta_1\cos(\Delta\phi)$

We obtain expressions for $\delta\theta$, $\delta\phi$, δh in terms of δa, δf, $\delta\theta_1$, $\delta\phi_1$, δh_1. These operations, although elementary in principle, turn out to be rather sophisticated in practice. Since $\xi = \theta_a - \theta_e$ and $\eta = (\phi_a - \phi_e)\sin\theta_e$, we can relate the variation in ellipsoidal coordinates to the variation in the vertical deflection; thus $d\xi = -d\theta_e$ and $d\eta = -(d\phi_e)\sin\theta_e$. Also, the geometric height h for a point P_1 on Earth from the reference ellipsoid, measured along the normal $\mathbf{n}_2$ to the ellipsoid, can be expressed approximately as the sum $h = N + H$, where H (the orthometric height) is the distance between P_1 on Earth and the geoid according to the plumb line $\mathbf{n}_1$ (normal to the geoid). Thus $dh = dN$.

In Table 1.2 we have expressed $\delta\xi$, $\delta\eta$, $\delta N/a$ in terms of $(\delta a/a)$, δf, $\delta\xi_1$, $\delta\eta_1$, and $(\delta N_1/a)$, and we have written $\Delta\phi = \phi - \phi_1$. All quantities therein are dimensionless coefficients.

Chapter II Hydrostatic Equilibrium and Earth Density Models

2.00 INTRODUCTION

This chapter is devoted to the study of the configuration of a rotating, self-gravitating body in hydrostatic equilibrium. By taking into consideration the internal constitution of the body, we reach the Clairaut equation, whose solution provides the rotational deformation of the body as a function of density. We have determined this equation up to terms in the third power of a small rotational parameter.

We use next the Poisson equation in conjunction with the Clairaut equation to deal with the problem of determining the change in the original density distribution because of rotation. We have devoted some time to describing certain procedures used in seismology to ascertain the density profile as well as the elastic parameters of Earth's interior. These methods are based on the elastic properties of the Earth; we believe, however, that they rightfully belong to this chapter because we make use of recent density profiles to integrate numerically the Clairaut equation. In so doing, we determine those parameters which are associated with the hydrostatic figure of Earth.

We close the chapter by discussing a method of celestial mechanics for determining Earth's exterior potential. Specifically, we examine the perturbations experienced by the Keplerian orbit of an artificial satellite due to the second, fourth, and sixth harmonics of Earth's potential and study how the numerical values of those harmonics can be determined from the tracking of artificial satellites. The main references to the subject of this chapter are some papers by Lanzano dealing with the Clairaut equation

(1962, 1969, 1974, and 1977); for the study of the density models one should consult a 1975 publication of Gilbert and Dziewonski; Bullen's book on seismology (1963) provides a very clear description of the early methods.

A final note on notation. We have covered in this chapter topics belonging to different subjects and have tried, as much as possible, to eliminate conflicts in notation. In this connection, let us mention, however, that the symbol μ has been used in this chapter to denote (1) the rigidity of a material (Lamé parameter), (2) the gravitational parameter of a body (Gm), and (3) the averaging operator of numerical analysis.

2.01 EQUILIBRIUM CONFIGURATION OF A ROTATING BODY OF ARBITRARY DENSITY

We consider a deformable, nonviscous body; we denote by ρ its density at any internal point, by p the pressure, and by ψ its gravitational potential plus the disturbing potential for all the external and inertial forces bearing on the body. The fundamental equation valid for hydrostatic equilibrium is

$$\nabla p = \rho \nabla \psi, \tag{1}$$

relating the gradient of the pressure to the gradient of the potential; it is obtained from Eulerian equations of motion by imposing the condition that there be no relative motion. The compatibility conditions of Eq. (1) are obtained by equating to each other the mixed second-order partial derivatives of p; using cartesian coordinates, they can be written as

$$\frac{\partial \rho}{\partial x}\bigg/\frac{\partial \psi}{\partial x} = \frac{\partial \rho}{\partial y}\bigg/\frac{\partial \psi}{\partial y} = \frac{\partial \rho}{\partial z}\bigg/\frac{\partial \psi}{\partial z}.$$

The above equations show that the normals to the surfaces $\rho =$ const. and $\psi =$ const. must be parallel and therefore that these surfaces must coincide. In other words, ψ is ultimately a function of ρ alone; due to Eq. (1), p must also be a function of ρ alone. We can then write generically $p = f(\rho)$ to represent the equation of state for the matter constituting the body under consideration. Any surface of constant density will be a surface of constant pressure and also an equipotential surface. This reasoning easily leads to the important fact that the equipotential surfaces $\psi =$ const. satisfy Eq. (1) and therefore must represent equilibrium configurations. In what follows we shall assume that (1) the disturbing potential is due solely to a rotation about an axis which remains fixed with respect to the rotating surface; (2) there exists symmetry with respect to the equatorial plane; (3) the equipotential surfaces are spheroidal surfaces, can be expanded into series of spherical harmonics, and depend continuously on a single parameter a, which we label the mean radius of the spheroid; and (4) the potential field of the rotating mass has no singular points, so that through each point within the mass there passes one and only one such equipotential spheroid.

We introduce a system of spherical coordinates $(O; r, \theta, \phi)$, with origin O at the centroid of the rotating mass; θ is the colatitude measured from the rotational axis, and ϕ is the longitude measured on the equatorial plane from an arbitrary axis. With respect to such a reference, the equipotential surfaces can be written as

$$r/a = 1 + \sum_{j=0}^{\infty} f_{2j}(a/a_1)P_{2j}(\cos\theta), \tag{2}$$

where the P are Legendre polynomials and a_1 is the mean radius of the outermost surface.

To obtain the unknown functions f_{2j} which represent the amplitudes of the various harmonics, we must take the following steps: (1) represent the potential ψ as a series of spherical harmonics; (2) replace within such an expansion of the potential the radius vector r by Eq. (2) and, by performing the requisite reductions, rearrange this expression into a new spherical harmonic representation of ψ; and (3) since this latter expansion for ψ must reduce to a constant, equate the coefficients of P_{2j} for $j = 1, 2, 3, \ldots$ to zero and the coefficient of P_0 to a constant.

This procedure leads to a linear differential equation first introduced by Clairaut and whose variable coefficients depend on the density profile within the body.

In effect, one should solve the broader problem of ascertaining the density distribution simultaneously with the deformation. This can be achieved by taking into account the Poisson equation

$$\nabla^2\psi(\rho) = -4\pi G\rho + 2\omega^2 \tag{3}$$

associated with the given potential. Here ∇^2 is the Laplace operator, G is the universal gravitational constant, and ω is the constant velocity of angular rotation. In the case of Earth, this last step can be avoided because the internal density distribution can be deduced from seismological evidence. In the case of the giant planets Jupiter and Saturn, astrophysical considerations will suggest a typical composition which will lead to a density profile.

Fundamental work on the Clairaut equation has been performed by De Sitter (1924) and Kopal (1960).

We will describe here a procedure for obtaining higher-order approximations for the deformations f in terms of the nondimensional parameter

$$q = \frac{\omega^2 a_1^3}{3Gm_1}, \tag{4}$$

where m_1 is the total mass of the body. For Earth, $q = 0.00115$ and can be considered a small parameter.

Preliminary order-of-magnitude calculations easily reveal that f_2 is of the first order in the rotational parameter q, that both f_0 and f_4 are of the second order, and that f_6 is of the third order, whereas f_8 and higher-order coefficients are at least of the fourth order in this parameter. If we limit ourselves to the third power of q, we shall therefore consider the equation of the equipotentials surfaces to be of the form

$$r/a = 1 + \sum_{j=0}^{3} \sum_{k=1}^{3} q^k f_{2j,k}(a/a_1) P_{2j}(\cos\theta), \tag{5}$$

or explicitly

$$\begin{aligned} r/a &= 1 + (f_{02}q^2 + f_{03}q^3)P_0 + (f_{21}q + f_{22}q^2 + f_{23}q^3)P_2 \\ &\quad + (f_{42}q^2 + f_{43}q^3)P_4 + f_{63}q^3 P_6 \\ &= 1 + f_0 P_0 + f_2 P_2 + f_4 P_4 + f_6 P_6. \end{aligned} \tag{6}$$

Our problem will then consist of determining the above eight coefficients.

Let us begin by providing some mathematical preliminaries for the purpose of covering in a systematic manner the various steps listed above.

The only disturbing potential envisaged here is the rotational potential; this can be expressed in terms of the parameter q and the Legendre polynomials as follows:

$$W = \tfrac{1}{3}\omega^2 r^2 \sin^2\theta = q(Gm_1/a_1^3)r^2(1 - P_2). \tag{7}$$

The gravitational potential at an interior point $Q(r, \theta, \phi)$, located on the equipotential surface of mean radius a, consists of two terms: (1) $V(r, \theta, \phi)$, which is due to the attraction on Q by the mass within this equipotential surface; and (2) $U(r, \theta, \phi)$, due to the attraction of the mass outside this equipotential surface and extending as far as the external boundary. Sometimes these two terms are referred to as the inner and outer contributions, respectively; otherwise, as the core and shell terms of the interior potential. It is well known (see, e.g., Tisserand, 1891), that

$$V = \sum_{j=0}^{\infty} r^{-(1+j)} V_j(\theta, \phi), \qquad U = \sum_{j=0}^{\infty} r^j U_j(\theta, \phi), \tag{8}$$

where

$$V_j = \frac{G}{3+j} \int_0^a \rho \left\{ \int_0^{2\pi} \int_0^{\pi} r^{3+j} P_j(\cos\gamma) \sin\bar{\theta}\, d\bar{\theta}\, d\bar{\phi} \right\}' d\bar{a}, \qquad j = 0, 1, 2, \ldots, \tag{9}$$

$$U_j = \frac{G}{2-j} \int_a^{a_1} \rho \left\{ \int_0^{2\pi} \int_0^{\pi} r^{2-j} P_j(\cos\gamma) \sin\bar{\theta}\, d\bar{\theta}\, d\bar{\phi} \right\}' d\bar{a}, \qquad j \neq 2.$$

Primes here denote derivatives with respect to the parameter a. The angle γ represents the angular separation between two vectors issued from the centroid O: one of these extends to the fixed point $Q(r, \theta, \phi)$ at which the potential is to be evaluated; the second vector terminates at the center $\bar{Q}(\bar{r}, \bar{\theta}, \bar{\phi})$ of the variable attracting element of mass. Clearly we must have

$$\cos\gamma = \cos\theta\cos\bar{\theta} + \sin\theta\sin\bar{\theta}\cos(\phi - \bar{\phi}).$$

The expression for U_j as given by Eq. (9) is indeterminate for $j = 2$ because the numerator

$$(r^{2-j})'$$

will vanish at the same time as the denominator $2 - j$. A simple application of the l'Hospital rule for indeterminate forms easily shows that

$$U_2 = G\int_a^{a_1} \rho\left\{\int_0^{2\pi}\int_0^{\pi} (\ln r)P_2(\cos\gamma)\sin\bar{\theta}\,d\bar{\theta}\,d\bar{\phi}\right\}' d\bar{a}, \qquad (10)$$

where the logarithm is to the base e.

In performing the integration with respect to the angular variables θ, ϕ, we must take into account the orthogonality conditions for spherical harmonics; these are represented by

$$\int_0^{2\pi}\int_0^{\pi} P_j(\cos\gamma)P_k(\cos\bar{\theta})\sin\bar{\theta}\,d\bar{\theta}\,d\bar{\phi} = \frac{4\pi}{1 + 2j}\,\delta_k^j P_j(\cos\theta), \qquad (11)$$

where the Kronecker deltas are defined as

$$\delta_k^j = \begin{cases} 0 & \text{if } j \neq k \\ 1 & \text{if } j = k. \end{cases}$$

Another problem encountered in various operations of simplification is how to express the product of two or three Legendre polynomials in terms of a linear combination of these polynomials. This can be achieved by repeated application of the Adams–Neumann formula, which will be given in Chapter III in this book. These products can be represented symbolically as

$$P_jP_k = \sum_h A_{jk}^h P_h, \qquad h = j + k, j + k - 2, \ldots, j - k,$$

$$P_jP_kP_r = \sum_h A_{jkr}^h P_h, \qquad h = j + k + r, j + k + r - 2, \ldots,$$

where, without loss of generality, we may assume that $j \geqslant k \geqslant r$. This simply means that A_{jkr}^h is the numerical coefficient that multiplies the P_h which appears as a term in the unique expansion of the product $P_jP_kP_r$ in Legendre polynomials; it must be construed as being equal to zero if no such term appears in the expansion. Thus, e.g., A_{20}^j is the Kronecker delta δ_{2j} since

$P_2P_0 = P_2$. This notation allows complete generality and compactness in the mathematical formulation of the results. It can be seen quite easily that the multiplication table required in what follows is

$$\begin{aligned}(P_2)^2 &= \tfrac{18}{35}P_4 + \tfrac{2}{7}P_2 + \tfrac{1}{5}P_0,\\ P_4P_2 &= \tfrac{5}{11}P_6 + \tfrac{20}{77}P_4 + \tfrac{2}{7}P_2,\\ (P_2)^3 &= \tfrac{18}{77}P_6 + \tfrac{108}{385}P_4 + \tfrac{3}{7}P_2 + \tfrac{2}{35}P_0.\end{aligned} \tag{12}$$

Using the previously introduced compact notation, we can write the generic nth power for the radius vector of the equipotential surfaces as

$$\begin{aligned}\frac{(r/a)^n - 1}{n} = \sum_{j=0}^{3} \Bigg\{ & q f_{2j,1} + q^2\left(f_{2j,2} + \frac{n-1}{2} A_{22}^{2j} f_{21}^2\right)\\ & + q^3\left[f_{2j,3} + (n-1)B_{2j} + \frac{(n-1)(n-2)}{6} A_{222}^{2j} f_{21}^3\right]\Bigg\} P_{2j},\end{aligned} \tag{13}$$

where

$$B_{2j} = \sum_{k=0}^{2} A_{2k,2}^{2j} f_{21} f_{2k,2} \tag{14}$$

and we have used a comma to separate the lower indices when they appear in literal form.

Let us also note that because of the well-known limit

$$\lim_{n\to 0} \frac{(r/a)^n - 1}{n} = \ln(r/a),$$

we can obtain the expansion of $\ln(r/a)$ simply by setting $n = 0$ in the right-hand side of Eq. (13). Because of the orthogonality conditions of the Legendre polynomials expressed by Eq. (11), it is evident that for a fixed value of j, the integral representations of V_j, U_j $(j \neq 2)$, and U_2 according to Eqs. (9) and (10) will be proportional to P_j and that their proportionality factors will depend on the coefficients of P_j appearing in the expansions of r^{j+3}, r^{2-j}, and $\ln(r/a)$, respectively.

Within the framework of the third-order approximation, any power of r will be limited to terms in P_0, P_2, P_4, and P_6 [see Eq. (13)]; this simply means that we have to take into account only those U_j and V_j for which $j = 0, 2, 4, 6$. Our calculations should be limited therefore to the following expression:

$$\psi = W + U_0 + r^2U_2 + r^4U_4 + r^6U_6 + \frac{V_0}{r} + \frac{V_2}{r^3} + \frac{V_4}{r^5} + \frac{V_6}{r^7}.$$

By using the expression for the various powers of r, Eq. (13), and the multiplication table for the Legendre polynomials [Eq. (12)], we can reduce the above potential to the form

$$\psi(a,\theta,q) = (\psi_{00} + \psi_{01}q + \psi_{02}q^2 + \psi_{03}q^3)P_0 + (\psi_{21}q + \psi_{22}q^2 + \psi_{23}q^3)P_2 + (\psi_{42}q^2 + \psi_{43}q^3)P_4 + \psi_{63}q^3P_6, \tag{15}$$

where the ψ with two subscripts are functions of a alone.

2.02 CLAIRAUT EQUATION EXPRESSED UP TO THIRD-ORDER TERMS

We proceed to ascertain the eight unknown functions

$$f_{02}, f_{03}; f_{21}, f_{22}, f_{23}; f_{42}, f_{43}; f_{63},$$

which define the hydrostatic deformation of a rotating earth.

The conservation of total mass within the distorted configuration allows us to express the functions f_{02}, f_{03} in terms of f_{21}, f_{22}. In fact, if we write the total mass as the integral

$$m_1 = m(a_1) = \tfrac{1}{3}\int_0^{a_1} \rho\left\{\int_0^{2\pi}\int_0^{\pi} r^3 \sin\theta\, d\theta\, d\phi\right\}' da,$$

and use Eq. (13) with $n = 3$ and $j = 0$ to eliminate r^3, we find

$$m_1 = 4\pi\int_0^{a_1} \rho a^2\, da + 4\pi q^2 \int_0^{a_1} \rho[a^3(f_{02} + A_{22}^0 f_{21}^2)]'\, da + 4\pi q^3 \int_0^{a_1} \rho[a^3(f_{03} + 2A_{22}^0 f_{21} f_{22} + \tfrac{1}{3}A_{222}^0 f_{21}^3)]'\, da.$$

Here ρ represents the density of the undistorted configuration, and no change in total mass will occur because of rotation; it follows then that the first term appearing in the right-hand side of the above equation should by itself represent the whole mass. The vanishing of the other two integrals yields

$$\begin{aligned} f_{02} &= -A_{22}^0 f_{21}^2 = -\tfrac{1}{5} f_{21}^2, \\ f_{03} &= -2A_{22}^0 f_{21} f_{22} - \tfrac{1}{3}A_{222}^0 f_{21}^3 = -\tfrac{2}{5} f_{21} f_{22} - \tfrac{2}{105} f_{21}^3. \end{aligned} \tag{16}$$

To solve for the other six unknowns we must impose the vanishing of the coefficients of P_2, P_4, and P_6 which appear in the expression of the total potential given by Eq. (15). We get six integral equations for the unknown functions, which can be conveniently grouped into three sets. Using the notation already introduced, we can write these equations in the following compact form:

$$\psi_{21}(a) = -(Gm_1/a_1)(a/a_1)^2 + (4\pi G/a)\left[-f_{21}\int_0^a \rho a^2\, da + (a^{-2}/5)\int_0^a \rho(a^5 f_{21})'\, da + (a^3/5)\int_a^{a_1} \rho f'_{21}\, da\right] = 0;$$

$$\begin{aligned}\psi_{2j,2} = {} & 2(Gm_1/a_1)(a/a_1)^2(A_{20}^{2j} - A_{22}^{2j})f_{21} \\ & + (4\pi G/a)\Big\{[a^{-2j}/(1+4j)]\int_0^a \rho(a^{3+2j}F_{2j,2})'\, da \\ & + [a^{1+2j}/(1+4j)]\int_a^{a_1} \rho(a^{2-2j}E_{2j,2})'\, da \\ & + \tfrac{1}{5}A_{22}^{2j}\left[2a^3\int_a^{a_1}\rho f'_{21}\, da - \frac{3}{a^2}\int_0^a \rho(a^5 f_{21})'\, da\right]f_{21} \\ & - (f_{2j,2} - A_{22}^{2j}f_{21}^2)\int_0^a \rho a^2\, da\Big\} = 0, \qquad j = 1, 2,\end{aligned}$$

where

$$\begin{aligned}E_{2j,2} &= f_{2j,2} + \tfrac{1}{2}(1-2j)A_{22}^{2j}f_{21}^2, \\ F_{2j,2} &= f_{2j,2} + (1+j)A_{22}^{2j}f_{21}^2;\end{aligned} \tag{17}$$

and

$$\begin{aligned}\psi_{2j,3} = {} & (Gm_1/a_1)(a/a_1)^2\left[2f_{2j,2} + A_{22}^{2j}f_{21}^2 - \sum_{k=0}^{2} A_{2k,2}^{2j}(2f_{2k,2} + A_{22}^{2k}f_{21}^2)\right] \\ & + (4\pi G/a)\Big\{[a^{-2j}/(1+4j)]\int_0^a \rho(a^{3+2j}F_{2j,3})'\, da \\ & + [a^{1+2j}/(1+4j)]\int_a^{a_1} \rho(a^{2-2j}E_{2j,3})'\, da \\ & + \tfrac{1}{5}A_{22}^{2j}\left[2a^3\int_a^{a_1}\rho E'_{22}\, da - \frac{3}{a^2}\int_0^a \rho(a^5F_{22})'\, da\right]f_{21} \\ & + \tfrac{1}{9}A_{42}^{2j}\left[4a^5\int_a^{a_1}\rho(a^{-2}E_{42})'\, da - \frac{5}{a^4}\int_0^a \rho(a^7F_{42})'\, da\right]f_{21} \\ & - \tfrac{3}{5}a^{-2}\sum_{k=0}^{2} A_{2k,2}^{2j}(f_{2k,2} - 2A_{22}^{2k}f_{21}^2)\int_0^a \rho(a^5f_{21})'\, da \\ & + \frac{a^3}{5}\sum_{k=0}^{2} A_{2k,2}^{2j}(2f_{2k,2} + A_{22}^{2k}f_{21}^2)\int_a^{a_1}\rho f'_{21}\, da \\ & - (f_{2j,3} - 2B_{2j} + A_{222}^{2j}f_{21}^3)\int_0^a \rho a^2\, da\Big\} = 0, \qquad j = 1, 2, 3,\end{aligned}$$

where

$$
\begin{aligned}
E_{2j,3} &= f_{2j,3} + (1-2j)B_{2j} - \tfrac{1}{3}j(1-2j)A_{222}^{2j}f_{21}^{3},\\
F_{2j,3} &= f_{2j,3} + 2(1+j)B_{2j} + \tfrac{1}{3}(1+j)(1+2j)A_{222}^{2j}f_{21}^{3},
\end{aligned}
\tag{18}
$$

and the B_{2j} have been defined by Eq. (14).

The first equation contains only f_{21}; the second set of two equations contains f_{22} and f_{42} and also lower-order terms in f_{21}; the third set of three equations contains f_{23}, f_{43}, and f_{63} and also lower-order terms in f_{21}, f_{22}, and f_{42}. It is then clear that these equations must be solved sequentially.

In each of these integral equations the six unknowns appear in two distinct ways: (1) outside the integral sign; and (2) under two types of integrals—one extending from 0 to a, which we have labeled the "core" integral, the other extending from a to a_1, which we have called the "shell" integral. Note also that both integrals cannot be eliminated simultaneously (i.e., by only one differentiation) since they are multiplied by different powers of the variable a. It is our purpose to transform this integral system of equations into an equivalent system of differential equations with boundary conditions; this operation necessitates three differentiations for each one of the equations of the first and second set and two differentiations for each equation of the third set. We must also get the boundary conditions expressed at the free surface, i.e., for $a = a_1$.

The operations can be visualized as follows: (1) We must isolate each one of the integrals and differentiate it with respect to a so that we can represent the second integral in terms of quantities free of the integral sign. These two differentiations must be performed for the first and second set so that we can replace the representation of both integrals into the second and third set; the differentiations serve also the purpose of obtaining the boundary conditions. (2) We must perform an additional differentiation to complete the reduction to a differential system.

The above procedure is simple in principle but constitutes a rather challenging analytic undertaking because it implies the handling and reduction of hundreds of literal terms. Following these steps we reach a differential system of the second order only because two differentiations are essentially required to eliminate the integral terms of a given order.

The result is the Clairaut equation, which we write as follows:

$$
a^2 f''_{2j,n} + 6D(f_{2j,n} + af'_{2j,n}) - 2j(2j+1)f_{2j,n} = R_{2j,n}. \tag{19}
$$

This equation holds for the choice of indices: $n = 1, j = 1$; $n = 2, j = 1, 2$; and $n = 3$, $j = 1, 2, 3$. The primes denote derivatives with respect to the variable a. Here

$$
D = \rho/\bar{\rho}, \tag{20}
$$

where

$$\bar{\rho}(a) = \frac{3}{a^3} \int_0^a \rho(a) a^2 \, da \tag{21}$$

is the mean density. The functions $R(a)$ which appear in the right-hand side depend on lower-order approximations; they have been found to be

$$R_{21} = 0; \tag{22}$$

$$\begin{aligned} R_{2j,2} = {} & A_{22}^{2j}\{(2 - 9D)\eta_{21}^2 + 2[12 - j(1 + 2j) - 9D]\eta_{21} \\ & + 3D(1 - j)(3 + 2j)\} f_{21}^2 + 12A_{20}^{2j} H(1 - D)(1 + \eta_{21}) f_{21}, \qquad j = 1, 2, \end{aligned} \tag{23}$$

where

$$\eta = af'/f \tag{24}$$

is the logarithmic derivative and

$$H(a) = \frac{3m_1}{4\pi a_1^3 \bar{\rho}(a)}, \tag{25}$$

so that $H(a_1) = 1$; and

$$\begin{aligned} R_{2j,3} = {} & 2 \sum_{k=1}^{2} A_{2k,2}^{2j}\{(2 - 9D)\eta_{21}\eta_{2k,2} \\ & + [3(1 - 3D) + 3k(1 + 2k) - j(1 + 2j)]\eta_{21} \\ & + [9(1 - D) + k(1 + 2k) - j(1 + 2j)]\eta_{2k,2} \\ & + 3D[k(1 + 2k) - j(1 + 2j)]\} f_{21} f_{2k,2} \\ & - A_{222}^{2j}\{3(2 - 5D)\eta_{21}^3 - 2[j(1 + 2j) - 33 + 9(3 + j)D]\eta_{21}^2 \\ & + [3(2j^2 - 11j - 38)D - 2j(1 + 2j) + 66]\eta_{21} \\ & - 2D(2j^3 - 3j^2 - 2j + 57) - 6j(1 + 2j) + 126\} f_{21}^3 \\ & + \sum_{k=1}^{2} A_{2k,2}^{2j} A_{22}^{2k}\{6(1 - 3D)\eta_{21}^3 + [100 - 8k(1 + 2k) - 9(7 + 2j)D]\eta_{21}^2 \\ & + 2[24 - k(1 + 2k) - 6(8 + 3j)D]\eta_{21} \\ & + 12(12 - k - 2k^2)(1 - D) + 3(3 + 2k)(1 - k)(1 + 2j)D\} f_{21}^3 \\ & + A_{20}^{2j} A_{22}^{0}[(2 - 15D)\eta_{21}^3 + 4(16 - 9D)\eta_{21}^2 \\ & - 3(4 + 35D)\eta_{21} + 132(1 - D)] f_{21}^3 \\ & + 6HA_{22}^{2j}[3\eta_{21}^2 + 6\eta_{21} - (1 - j)(3 + 2j)](1 - D) f_{21}^2 \\ & + 12H(1 - D)(1 + \eta_{2j,2}) f_{2j,2} + 24H^2(1 - D) A_{20}^{2j}(1 + \eta_{21}) f_{21}, \\ & \qquad j = 1, 2, 3. \end{aligned} \tag{26}$$

We can then solve each one of the six Clairaut equations to determine sequentially $f_{21}; f_{22}, f_{42}; f_{23}, f_{43}, f_{63}$. Table 2.1 gives explicitly the boundary

Table 2.1

Boundary Conditions That the Solutions to the Clairaut Equations Must Satisfy at the Outermost Surface

$$2f_{21} + a_1 f'_{21} + 5 = 0$$

$$2f_{22} + a_1 f'_{22} = \tfrac{2}{7}(6 + 3\eta_{21} + \eta_{21}^2)f_{21}^2 + 2(5 + \eta_{21})f_{21}$$

$$4f_{42} + a_1 f'_{42} = \tfrac{18}{35}(6 + 5\eta_{21} + \eta_{21}^2)f_{21}^2$$

$$\begin{aligned}2f_{23} + a_1 f'_{23} = {} & 2(5 + \eta_{22})f_{22} - \tfrac{2}{7}(15 + 6\eta_{21} + 2\eta_{21}^2)f_{21}^2 \\ & + \tfrac{2}{7}(12 + 3\eta_{21} + 3\eta_{22} + 2\eta_{21}\eta_{22})f_{21}f_{22} \\ & + \tfrac{2}{7}(26 + 3\eta_{21} + 3\eta_{42} + 2\eta_{21}\eta_{42})f_{21}f_{42} \\ & - \tfrac{4}{35}(38 + 30\eta_{21} + 15\eta_{21}^2 + 2\eta_{21}^3)f_{21}^3\end{aligned}$$

$$\begin{aligned}4f_{43} + a_1 f'_{43} = {} & 2(7 + \eta_{42})f_{42} - \tfrac{18}{35}(21 + 10\eta_{21} + 2\eta_{21}^2)f_{21}^2 \\ & + \tfrac{18}{35}(12 + 5\eta_{21} + 5\eta_{22} + 2\eta_{21}\eta_{22})f_{21}f_{22} \\ & + \tfrac{20}{77}(26 + 5\eta_{21} + 5\eta_{42} + 2\eta_{21}\eta_{42})f_{21}f_{42} \\ & - \tfrac{36}{385}(54 + 44\eta_{21} + 18\eta_{21}^2 + 3\eta_{21}^3)f_{21}^3\end{aligned}$$

$$6f_{63} + a_1 f'_{63} = \tfrac{5}{11}(26 + 7\eta_{21} + 7\eta_{42} + 2\eta_{21}\eta_{42})f_{21}f_{42} - \tfrac{18}{77}(2 + \eta_{21})(12 + 6\eta_{21} + \eta_{21}^2)f_{21}^3$$

Table 2.2

Explicit Expressions for the Functions Appearing in the Right-Hand Side of the Clairaut Equations

$$R_{21} = 0$$

$$R_{22} = \tfrac{2}{7}[(2 - 9D)\eta_{21} + 18(1 - D)]\eta_{21}f_{21}^2 + 12H(1 - D)(1 + \eta_{21})f_{21}$$

$$R_{42} = \tfrac{18}{35}[(2 - 9D)\eta_{21}(2 + \eta_{21}) - 21D]f_{21}^2$$

$$\begin{aligned}R_{43} = {} & \tfrac{36}{35}[(2 - 9D)(\eta_{21}\eta_{22} + \eta_{21} + \eta_{22}) - 21D]f_{21}f_{22} \\ & + \tfrac{40}{77}[(2 - 9D)\eta_{21}\eta_{42} + (23 - 9D)\eta_{21} + 9(1 - D)\eta_{42}]f_{21}f_{42} \\ & - \tfrac{36}{385}[9D\eta_{21}^3 - (10 - 27D)\eta_{21}^2 + 4(8 + 9D)\eta_{21} - 6(1 - 2D)]f_{21}^3 \\ & + 12H(1 - D)(1 + \eta_{42})f_{42} + \tfrac{108}{35}H(3\eta_{21}^2 + 6\eta_{21} + 7)(1 - D)f_{21}^2\end{aligned}$$

$$\begin{aligned}R_{23} = {} & \tfrac{4}{7}[(2 - 9D)\eta_{21}\eta_{22} + 9(1 - D)(\eta_{21} + \eta_{22})]f_{21}f_{22} \\ & + \tfrac{4}{7}[(2 - 9D)\eta_{21}\eta_{42} + 3(10 - 3D)\eta_{21} + (16 - 9D)\eta_{42} + 21D]f_{21}f_{42} \\ & - \tfrac{4}{35}[(7 + 6D)\eta_{21}^3 + 3(11 - 15D)\eta_{21}^2 + 9(20 - 9D)\eta_{21} + 3(22 + 5D)]f_{21}^3 \\ & + 12H(1 - D)(1 + \eta_{22})f_{22} + \tfrac{36}{7}H(1 - D)(2 + \eta_{21})\eta_{21}f_{21}^2 \\ & + 24H^2(1 - D)(1 + \eta_{21})f_{21}\end{aligned}$$

$$\begin{aligned}R_{63} = {} & \tfrac{10}{11}[(2 - 9D)\eta_{21}\eta_{42} + 3(4 - 3D)\eta_{21} - (2 + 9D)\eta_{42} - 33D]f_{21}f_{42} \\ & + \tfrac{18}{77}[-3D\eta_{21}^3 - (4 + 9D)\eta_{21}^2 + (4 - 45D)\eta_{21} + 24 - 15D]f_{21}^3\end{aligned}$$

conditions at the free surface, $a = a_1$. Table 2.2 provides explicit representations for R_{21}; R_{22}, R_{42}; R_{23}, R_{43}, R_{63}, Since the deformations must vanish at the center of the configuration, we must add the other boundary conditions

$$f_{2j,n}(0) = 0. \tag{27}$$

More detailed information about the stepwise elimination procedure can be obtained in Lanzano (1974). One can easily verify that the logarithmic derivative η of the solution to the Clairaut equation, as defined by Eq. (24), satisfies the equation

$$f[a\eta' + 6D(1 + \eta) - \eta(1 - \eta) - 2j(2j + 1)] = R. \tag{28}$$

This is known as the Radau equation and holds for the same choice of subscripts as the original Clairaut equation.

2.03 EXTERIOR POTENTIAL AND MOMENTS OF INERTIA

We intend to deal here with two immediate applications of the previous analytical formulation: the evaluation of the exterior potential and of the moments of inertia pertaining to the distorted configuration.

Up to third-order terms, the potential generated at an exterior point by the distorted configuration can be written as

$$V(r, \theta; a_1) = \frac{Gm_1}{r}\left[1 + \sum_{j=1}^{3} K_{2j}(a_1)\left(\frac{a_1}{r}\right)^{2j} P_{2j}(\cos\theta)\right], \tag{29}$$

where

$$K_{2j}(a_1) = \sum_{k=1}^{3} q^k K_{2j,k}(a_1), \qquad j = 1, 2, 3, \tag{30}$$

and

$$K_{2j,k}(a_1) = \frac{4\pi}{m_1}\frac{a_1^{-2j}}{1 + 4j}\int_0^{a_1} \rho(a^{3+2j}F_{2j,k})'\, da. \tag{31}$$

It is understood that

$$F_{2j,1} = f_{2j,1} \tag{32}$$

and that the F for $k = 2, 3$ are given by Eqs. (17) and (18). The integrals that constitute Eq. (31) are among those that had to be evaluated in order to express the Clairaut equation in its differential form; these integrals are therefore available to us.

Using those results and the boundary conditions expressed by Table 2.1, one can verify that

$$\begin{aligned}
K_{21}(a_1) &= -\frac{3-\eta_{21}(a_1)}{2+\eta_{21}(a_1)},\\
K_{22}(a_1) &= -\frac{20}{7}\frac{[6+15\eta_{21}(a_1)+\eta_{21}^2(a_1)-7\eta_{22}(a_1)-\eta_{21}(a_1)\eta_{22}(a_1)]}{[2+\eta_{21}(a_1)]^2[2+\eta_{22}(a_1)]},\\
K_{42}(a_1) &= \frac{90}{7}\frac{1}{[2+\eta_{21}(a_1)]}\left[\frac{3+\eta_{21}(a_1)}{4+\eta_{42}(a_1)}-\frac{\eta_{21}(a_1)}{2+\eta_{21}(a_1)}\right],\\
K_{2j,3}(a_1) &= f_{2j,3}(a_1)-2f_{2j,2}(a_1)+A_{22}^{2j}[3+2\eta_{21}(a_1)]\,f_{21}^2(a_1)\\
&\quad-\sum_{k=1}^{2}A_{2k,2}^{2j}[\eta_{21}(a_1)+\eta_{2k,2}(a_1)]f_{21}(a_1)f_{2k,2}(a_1)\\
&\quad+\{A_{222}^{2j}[3+\eta_{21}(a_1)+\eta_{21}^2(a_1)]+\tfrac{3}{5}A_{02}^{2j}\eta_{21}(a_1)\\
&\quad-\tfrac{1}{5}A_{22}^{2j}[5+2\eta_{21}(a_1)+\eta_{21}^2(a_1)]\}f_{21}^3(a_1), \qquad j=1,2,3.
\end{aligned} \tag{33}$$

These formulas establish a link between the rotational deformations of the body and the coefficients of its exterior potential. Should one set—say, the K—become available, then the other set—i.e., the f and η—could be ascertained by numerical procedures. In the last section of this chapter, we shall consider an application of celestial mechanics whereby we can ascertain the K-coefficients by measuring the perturbations exhibited by the orbit of an artificial satellite of real Earth as compared with the ideal case of a Keplerian orbit. The moment of inertia of the total distorted configuration with respect to any barycentric axis lying in the equatorial plane can be written as

$$I_e = \frac{2\pi}{5}\int_0^{a_1}\rho\left\{\int_0^{\pi}(r^5/3)[2+P_2(\cos\theta)]\sin\theta\,d\theta\right\}'da,$$

and the moment of inertia with respect to the polar (i.e., rotational) axis is given by

$$I_p = \frac{2\pi}{5}\int_0^{a_1}\rho\left\{\int_0^{\pi}(2r^5/3)[1-P_2(\cos\theta)]\sin\theta\,d\theta\right\}'da.$$

Primes, as usual, denote derivatives with respect to the mean radius a. The square of the distance of the variable element of mass to the axis in question has introduced the factors

$$\tfrac{1}{2}(1+\cos^2\theta)=\tfrac{1}{3}(2+P_2), \qquad \sin^2\theta=\tfrac{2}{3}(1-P_2). \tag{34}$$

These were expressed in terms of the Legendre polynomials in order that we could take advantage of the orthogonality conditions of these polynomials for integration purposes. Using the results of the Clairaut equation, we can write

$$\begin{aligned} I_e = \frac{4\pi}{15} \int_0^{a_1} \rho a^4 \{ & 10 + q(5 + \eta_{21}) f_{21} \\ & + q^2[(5 + \eta_{22}) f_{22} + \tfrac{18}{7}(5 + 2\eta_{21}) f_{21}^2] \\ & + q^3[(5 + \eta_{23}) f_{23} + \tfrac{36}{7}(5 + \eta_{21} + \eta_{22}) f_{21} f_{22} \\ & + \tfrac{8}{7}(5 + \eta_{21} + \eta_{42}) f_{21} f_{42} + \tfrac{106}{105}(5 + 3\eta_{21}) f_{21}^3]\} \, da. \end{aligned} \tag{35}$$

Similarly,

$$\begin{aligned} I_e - I_p &= \frac{2\pi}{5} \int_0^{a_1} \rho \left\{ \int_0^{\pi} r^5 P_2(\cos\theta) \sin\theta \, d\theta \right\}' da \\ &= \frac{4\pi}{5} q \int_0^{a_1} \rho a^4 \{(5 + \eta_{21}) f_{21} + q[(5 + \eta_{22}) f_{22} + \tfrac{4}{7}(5 + 2\eta_{21}) f_{21}^2] \\ &\quad + q^2[(5 + \eta_{23}) f_{23} + \tfrac{8}{7}(5 + \eta_{21} + \eta_{22}) f_{21} f_{22} \\ &\quad + \tfrac{8}{7}(5 + \eta_{21} + \eta_{42}) f_{21} f_{42} + \tfrac{2}{35}(5 + 3\eta_{21}) f_{21}^3]\} \, da. \end{aligned} \tag{36}$$

Note that this difference vanishes with q. It is to be understood that in this formulation the effective evaluation of both the K and the I can be undertaken only if the density $\rho(a)$ is known as a function of the mean radius a.

2.04 DENSITY DISTORTION GENERATED BY ROTATION

We utilize the results obtained in the previous sections to ascertain the distortion experienced by the original density distribution, which now we label $\rho_0(a)$, due to rotation. Because of the rotational motion, the original spherical shape of the body has been distorted into the spheroidal shape and the density becomes a function not only of the mean radius a but also of the colatitude θ:

$$\rho(a, \theta; q) = \rho_0(a) + \rho_1(a, \theta) q + \rho_2(a, \theta) q^2 + \cdots. \tag{37}$$

To solve for the unknown functions $\rho_n(a, \theta)$ for $n \geqslant 1$, knowing only the given $\rho_0(a)$, we must make use of the Poisson equation

$$\nabla^2 \psi(\rho) = -4\pi G \rho + 2\omega^2,$$

where we have emphasized the fact that the total potential ψ is essentially a function of ρ. This causes the Poisson equation to be in effect an integral equation since the unknown function ρ appears within the total potential under integral signs. This situation can be remedied, however, if we choose to approximate the Laplace operator ∇^2 by using the undistorted density $\rho_0(a)$ under those integral signs. We can then solve the equation

$$\nabla^2\psi(\rho_0) = -4\pi G\rho + 2\omega^2 \tag{38}$$

for ρ. This operation presupposes that the total potential for the distorted configuration is known; it is tantamount to knowing the distortion coefficients $f_{2j}(a)$ by having solved the Clairaut equation. We can considerably simplify computations by making use of the equations of the equipotential surfaces, which we write as $r = r(a, \theta; q)$. Upon each of these surfaces, the constant value of the potential can be simply expressed as

$$\psi(a, q) = \psi_{00}(a) + \psi_{01}(a)q + \psi_{02}(a)q^2 + \cdots. \tag{39}$$

This is so because in order to obtain the equation of the equipotential surfaces, we had to equate to zero the coefficients of the various harmonics appearing in the expression of the total potential [see Eq. (15)]. By using the results and the notation of the previous sections, we can easily show that

$$\begin{aligned} \psi_{00} &= 4\pi G\left[\tfrac{1}{3}a^2\bar{\rho}_0 + \int_a^{a_1} \rho_0 a\, da\right], \\ \psi_{01} &= \tfrac{1}{3}(4\pi G)a^2\bar{\rho}_0(a_1), \\ \psi_{02} &= \tfrac{1}{5}(4\pi G)\left[\tfrac{1}{3}a^2(1+\eta_{21})f_{21}^2\bar{\rho}_0 - \int_a^{a_1}(1+\eta_{21})f_{21}^2\rho_0 a\, da\right], \end{aligned} \tag{40}$$

where $\bar{\rho}_0(a_1) = 3m_1/4\pi a_1^3$ is the mean density of the whole mass.

Let us now switch from the orthogonal system of spherical coordinates (r, θ, ϕ) to the nonorthogonal system (a, θ, ϕ), where a is the mean radius that characterizes each equipotential surface. This can be achieved by replacing the family of spherical surfaces $r = \text{const.}$ with the family of spheroidal surfaces $r(a, \theta; q) = \text{const.}$ as one set of fundamental surfaces; the other two families of fundamental surfaces remain unchanged. From the transformation of coordinates

$$\begin{aligned} x &= r(a, \theta; q)\sin\theta\cos\phi, \\ y &= r(a, \theta; q)\sin\theta\sin\phi, \\ z &= r(a, \theta; q)\cos\theta, \end{aligned}$$

we can represent the element of arc between the two infinitesimally close points as the following quadratic form:

$$(ds)^2 = (dx)^2 + (dy)^2 + (dz)^2 = g_{ij}\,dx^i\,dx^j$$
$$= (r_a)^2(da)^2 + [r^2 + (r_\theta)^2](d\theta)^2 + (r\sin\theta)^2(d\phi)^2 + 2r_a r_\theta\,da\,d\theta. \qquad (41)$$

The subscripts a and θ are used here to denote partial derivatives with respect to those two variables; also the summation convention is used for the repeated indices in the tensorial representation of the element of arc. The appearance of the mixed term $(\cdots)\,da\,d\theta$ is indicative of the fact that two fundamental reference surfaces of our new system, specifically, the equipotential surfaces $r(a, \theta, q) = \text{const.}$ and the cones $\theta = \text{const.}$, are not orthogonal.

The determinant of the above quadratic form is

$$g = \begin{vmatrix} g_{11} & g_{12} & g_{13} \\ g_{21} & g_{22} & g_{23} \\ g_{31} & g_{32} & g_{33} \end{vmatrix} = \begin{vmatrix} (r_a)^2 & r_a r_\theta & 0 \\ r_a r_\theta & r^2 + (r_\theta)^2 & 0 \\ 0 & 0 & r^2\sin^2\theta \end{vmatrix}$$
$$= (r^2 r_a \sin\theta)^2. \qquad (42)$$

Identifying a by x^1, θ by x^2, and ϕ by x^3, we can write the Laplacian, using the tensorial notation, as follows:

$$\nabla^2\psi = \frac{1}{\sqrt{g}}\frac{\partial}{\partial x^i}\left(\sqrt{g}g^{ij}\frac{\partial\psi}{\partial x^j}\right), \qquad (43)$$

where the g^{ij} are the cofactors of the corresponding g_{ij} within the above determinant divided by the value of g.

Since ψ is a function of a alone and we have set $a = x^1$,

$$\nabla^2\psi = \frac{1}{\sqrt{g}}\frac{\partial}{\partial x^i}\left(\sqrt{g}g^{i1}\frac{\partial\psi}{\partial x^1}\right). \qquad (44)$$

We can easily verify from Eq. (42) that

$$\sqrt{g}g^{11} = \frac{r^2 + (r_\theta)^2}{r_a}\sin\theta, \qquad \sqrt{g}g^{21} = -r_\theta\sin\theta, \qquad \sqrt{g}g^{31} = 0. \qquad (45)$$

If we introduce the notation $v = \cos\theta$ and use Eqs. (38), (44), (45), and (42), we realize that the density distortion can be evaluated from the relation

$$-4\pi G\rho + 2\omega^2 = \frac{1}{r^2 r_a}\left\{\frac{\partial}{\partial a}\left[\frac{r^2 + (1 - v^2)(r_v)^2}{r_a}\frac{\partial\psi}{\partial a}\right]\right.$$
$$\left. - \frac{\partial}{\partial v}[(1 - v^2)r_v]\frac{\partial\psi}{\partial a}\right\}. \qquad (46)$$

Here again the subscripts a and v denote partial derivatives; $\psi(a)$ is given by Eqs. (39) and (40), where we must note that the known density $\rho_0(a)$ has been utilized.

We must expand the right-hand side of Eq. (46) in powers of q and then equate equal powers of q appearing in both sides of the equation to obtain $\rho_1(a, \theta)$ and $\rho_2(a, \theta)$. In this connection, we must note that

$$2\omega^2 = 6Gm_1q/a_1^3.$$

Our next task is to elaborate on the operator appearing in the right-hand side of Eq. (46) and show how its actual evaluation depends on the various literal expansions already encountered.

Any power of the radius vector r of the equipotential surfaces is obtainable from Eq. (13). A simple argument shows next that r_a is obtainable from Eq. (5), which represents the radius vector of the equipotential surfaces, by replacing f in it by $(1 + \eta)f$. Thus we get

$$r_a = \frac{\partial r}{\partial a} = 1 + \sum_{k=1}^{3} \sum_{j=0}^{3} q^k[(1 + \eta_{2j,k})f_{2j,k}]P_{2j}.$$

We obtain $(r_a)^{-1}$ by using Eq. (13) for $n = -1$, provided we have duly performed therein the replacement just mentioned. We use next the equation satisfied by the Legendre polynomials,

$$\frac{\partial}{\partial v}\left[(1 - v^2)\frac{\partial P_n}{\partial v}\right] = -n(n + 1)P_n, \tag{47}$$

to represent the following expansions:

$$-\frac{\partial}{\partial v}\left[(1 - v^2)\frac{\partial r}{\partial v}\right] = 2aq[3f_{21}P_2 + (3f_{22}P_2 + 10f_{42}P_4)q],$$

$$(1 - v^2)\left(\frac{\partial r}{\partial v}\right)^2 = 2a^2q^2(\tfrac{3}{5}f_{21}^2 + \tfrac{3}{7}f_{21}^2P_2 - \tfrac{36}{35}f_{21}^2P_4).$$

We use Eqs. (39) and (40) to obtain $\partial\psi/\partial a$; in doing so, we must express the derivatives of the f by means of the η, and the derivatives of the η shall be eliminated by means of the Radau equation (28). We use also Eq. (21) to obtain

$$\bar{\rho}_0' = 3(\rho_0 - \bar{\rho}_0)a^{-1},$$

which in turn can be used to eliminate $\bar{\rho}_0'$. The result of the previous operations is

$$(1/a)\psi_a = \tfrac{1}{3}(4\pi G)[-\bar{\rho}_0 + 2\bar{\rho}_0(a_1)q + \tfrac{1}{5}(5 + 2\eta_{21} + \eta_{21}^2)f_{21}^2\bar{\rho}_0q^2],$$

whose dependence on the density is only through the mean density $\bar{\rho}_0$.

After performing another differentiation with respect to a and the requisite series multiplications according to Eq. (46), which implies considerable simplifications and grouping of literal terms, we reach the result

$$\begin{aligned} \tfrac{1}{2}\rho_1(a,\theta) &\equiv -f_{21}\bar{\rho}_0 P_2, \\ \tfrac{1}{2}\rho_2(a,\theta) &\equiv \tfrac{1}{5}(3-\eta_{21})f_{21}^2\bar{\rho}_0 - \tfrac{3}{5}f_{21}^2\rho_0 \\ &\quad + [2f_{21}\bar{\rho}_0(a_1) + (\tfrac{6}{7}f_{21}^2 - f_{22})\bar{\rho}_0 - \tfrac{3}{7}f_{21}^2\rho_0]P_2 \\ &\quad + \{\tfrac{36}{35}f_{21}^2\rho_0 + [\tfrac{6}{35}(9+7\eta_{21})f_{21}^2 - \tfrac{10}{3}f_{42}]\bar{\rho}_0\}P_4. \end{aligned} \tag{48}$$

A more detailed description of the various operations is made in a paper by Lanzano (1975), where also the third-order approximation $\rho_3(a,\theta)$ has been obtained.

2.05 CONSIDERATIONS ON EARTH DENSITY MODELS

We plan to integrate the Clairaut equation in order to obtain the ellipticity of the family of equipotential surfaces as well as the rotational deformation of the outermost equilibrium surface. This numerical integration provides also results pertaining to Earth's exterior potential which will be compared with the data available from satellite geodesy. For the purpose of integration, a density profile $\rho_0(r)$ must be provided. In this as well as in future chapters, we shall make use of recent Earth density models obtainable from seismological considerations.

Although the subject of ascertaining the density distribution via seismological methods is strictly related to the elastic properties of Earth, it seems appropriate at this stage to provide a discussion of the problem to enable the reader to understand its merits and limitations so that he will be able to make a proper assessment of the validity of the numerical results.

Jeffreys and Bullen have pioneered work in which they used the fundamental fact that the travel times of bodily waves (P and S seismic waves) do not depend on the source mechanism of the earthquake but are functionals of the structure alone.

From readings of seismograms, one can derive empirical relations between the travel time T and the traveled angular distance Δ from the epicenter to the recording station, valid for various families of bodily waves. The matter is strictly related to the problem of ascertaining the parameters of an earthquake, i.e., time of origin, coordinates of the epicenter, and depth of the focus, by using a wide network of seismic stations. The set of data, refined by

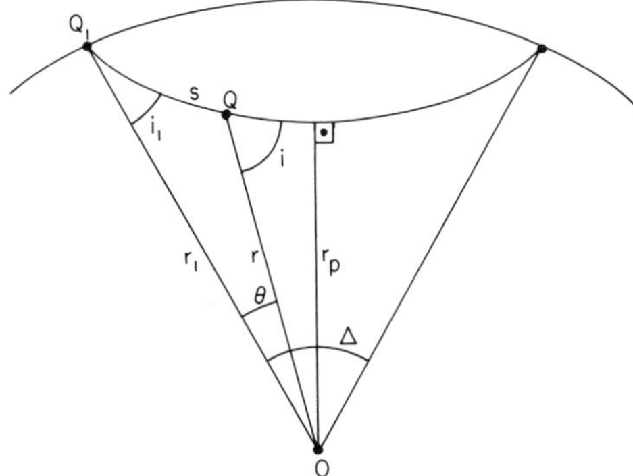

Fig. 2.1 Parameters pertaining to the propagation of seismic waves through a spherically stratified earth.

least squares methods, ultimately gives rise to the so-called Jeffreys–Bullen (J–B) travel time tables. It is possible therefore to ascertain T experimentally as a function of Δ. Further progress can be achieved by making recourse to the elementary theory of propagation of seismic waves through a spherically stratified Earth, see Fig. 2.1.

Using the law of refraction for waves through inhomogeneous media, one can easily establish the fact that the quantity

$$p = \frac{r \sin i}{v} = \eta \sin i \tag{49}$$

remains the same for every point of a given ray. Here v is the propagation velocity of the wave and $\eta = r/v$. The constant p is called the parameter of the ray. In particular, at the point of deepest penetration, which corresponds to the value r_p of the radius, one must have $\sin i = 1$, and then

$$p = \frac{r_p}{v_p} = \eta_p. \tag{50}$$

Next, by considering two infinitesimally close rays, an elementary argument establishes the property that the parameter p is the rate of change of T with respect to Δ, i.e.,

$$p = dT/d\Delta. \tag{51}$$

In other words, the slope of the graph $T(\Delta)$ is equal to p; we can then express Δ as a function of p, thereby obtaining $\Delta(p)$, albeit in graphical or tabular form.

Let us now refer the points of the ray to a polar coordinate system as shown in Fig. 2.1. The expression for the arc length is

$$(ds)^2 = (dr)^2 + r^2(d\theta)^2;$$

recalling then that $\sin i = r\,d\theta/ds$, we can write

$$p = \frac{r^2}{v}\frac{d\theta}{ds}.$$

If we eliminate ds in the two equations above, we reach the following expression for the traveled angular distance or epicentral distance:

$$\Delta = 2\int_{r_p}^{r_1} d\theta = 2\int_{r_p}^{r_1} \frac{p}{r}(\eta^2 - p^2)^{-1/2}\,dr. \tag{52}$$

Here r_1 denotes the radius of the outermost surface and r_p the radius of deepest penetration for the ray under consideration. In a similar fashion, if we eliminate the $d\theta$ in the same two previous equations, we find

$$T = 2\int_{r_p}^{r_1} \frac{ds}{v} = 2\int_{r_p}^{r_1} \frac{\eta^2}{r}(\eta^2 - p^2)^{-1/2}\,dr. \tag{53}$$

Simple algebraic manipulation of Eqs. (52) and (53) reveals that

$$T = p\Delta + 2\int_{r_p}^{r_1} \frac{1}{r}(\eta^2 - p^2)^{1/2}\,dr. \tag{54}$$

The propagation velocity v can be shown to be a slowly increasing function of the depth z; its value undergoes finite jumps because of the density discontinuity between adjacent layers. Accepting this experimental fact, we can say that $\eta = r/v$ is a decreasing function of depth. The most expeditious and advantageous way from a seismological viewpoint to corroborate the behavior of Δ as a function of p is to introduce a new variable ξ, given by

$$\xi/2 = \frac{d(\ln r)}{d(\ln \eta)} = \frac{\eta}{r}\frac{dr}{d\eta},$$

and to study the derivative

$$\frac{d\Delta}{dp} = -\xi_1(\eta_1^2 - p^2)^{-1/2} + \int_{\xi_p}^{\xi_1} (\eta^2 - p^2)^{-1/2}\,d\xi,$$

with the same proviso as to the significance of the subscripts 1 and p as has been mentioned. To obtain the solution of the fundamental problem of expressing the propagation velocity as a function of the penetration depth, let us revert to Eq. (52) and rewrite it as

$$\Delta(p) = \int_{\eta_p}^{\eta_1} \frac{2p}{r} (\eta^2 - p^2)^{-1/2} \frac{dr}{d\eta} \, d\eta, \tag{55}$$

in which we emphasize the fact that η is a function of r and that $\Delta(p)$ is a known function of p. This expression can be interpreted as an integral equation of the Abel type; its solution will provide η, and ultimately v, as a function of r.

For the purpose of solving Eq. (55), let us apply to both sides of it the integral operator

$$\int_{\eta_2}^{\eta_1} (p^2 - \eta_2^2)^{-1/2} \, dp,$$

where we are assuming that $\eta_1 > \eta_2 \geqslant \eta_p$ because of the decreasing character of η with depth; this also translates into the same type of inequality, $r_1 > r_2 \geqslant r_p$. The interpretation of the above operation is that we are integrating the given expression along a bundle of rays whose deepest penetrations vary between the outermost surface, corresponding to η_1, and a fixed depth r_2, corresponding to η_2.

Carrying out the proposed operation gives

$$\int_{\eta_2}^{\eta_1} \Delta(p)(p^2 - \eta_2^2)^{-1/2} \, dp = \int_{\eta_2}^{\eta_1} \left\{ \int_{\eta_p}^{\eta_1} \frac{2p}{r} (p^2 - \eta_2^2)^{-1/2} (\eta^2 - p^2)^{-1/2} \frac{dr}{d\eta} \, d\eta \right\} dp. \tag{56}$$

In order to evaluate this expression, we shall interchange the order of integration of the double integral appearing on its right-hand side. For this purpose, let us remark that in the (ηp)-plane the domain of integration is the triangular region bounded by the three lines $\eta = p$, $p = \eta_2$, and $\eta = \eta_1$; see Fig. 2.2. This is so because p must be chosen between η_2 and η_1; also, in integrating with respect to η, this variable shall always remain larger or equal to p. This suggests that the above double integral can also be written as

$$\int_{\eta_2}^{\eta_1} \left\{ \int_{\eta_2}^{\eta} 2p(p^2 - \eta_2^2)^{-1/2} (\eta^2 - p^2)^{-1/2} \, dp \right\} \frac{1}{r} \frac{dr}{d\eta} \, d\eta. \tag{57}$$

The inner integral in Eq. (57) can be evaluated by elementary methods. In fact, the transformation of variables

$$p^2 = \frac{\eta^2 + \eta_2^2 z^2}{1 + z^2}$$

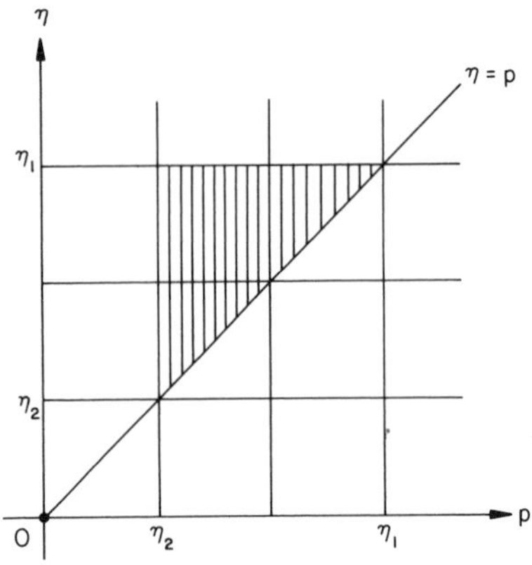

Fig. 2.2 Integration domain for the double integral which was used in solving for the propagation velocity of a seismic wave as a function of the penetration depth.

will reduce it to $2\int_0^\infty (dz)/(1+z^2) = \pi$; thus the double integral appearing in Eq. (57) reduces to

$$\pi \int_{r_2}^{r_1} \frac{dr}{r} = \pi \ln(r_1/r_2).$$

If we integrate the left-hand side of Eq. (56) by parts, we obtain the expression

$$[\Delta(p)\cosh^{-1}(p/\eta_2)]_{\eta_2}^{\eta_1} - \int_{\Delta_2}^{\Delta_1} \cosh^{-1}(p/\eta_2)\, d\Delta.$$

Now $\cosh^{-1}(1) = 0$; also, $\Delta(\eta_1) = \Delta_1 = 0$ since the traveled angular distance of the ray whose deepest penetration is r_1, i.e., the outermost surface, must necessarily be equal to zero. Equation (56) now simplifies to

$$\int_0^{\Delta_2} \cosh^{-1}(p/\eta_2)\, d\Delta = \pi \ln(r_1/r_2). \tag{58}$$

The expression on the left can be numerically integrated since we know Δ as a function of p; η_2 is also known since it equals the value of $dT/d\Delta$ evaluated at Δ_2. Thus r_2 can be expressed in terms of Δ_2 and η_2, i.e., in terms of v_2. Due to the arbitrariness in the choice of η_2, the above procedure applies to the general case.

Once we have ascertained the velocities $v_p = \alpha$ and $v_s = \beta$ of the P and S seismic waves as a function of depth z, we can use the well-known formulas

$$\alpha^2 - \tfrac{4}{3}\beta^2 = k/\rho_0 = \phi, \qquad \beta^2 = \mu/\rho_0 \tag{59}$$

to establish relations involving the elastic parameters of the material. Here k stands for the incompressibility of the material and can be represented as

$$k = \rho_0(dp/d\rho_0),$$

where p is the pressure and μ is the rigidity of the material. Assuming, for simplicity, a hydrostatic variation of pressure with depth, we can then represent the variation of density with depth as

$$\frac{d\rho_0}{dz} = \frac{d\rho_0}{dp}\frac{dp}{dz} = \frac{\rho_0}{k}(\rho_0 g_0) = \frac{Gm\rho_0}{r^2\phi}. \tag{60}$$

We shall also take advantage of the mass variation with radial distance, i.e., of the relation

$$dm = 4\pi r^2 \rho_0\, dr. \tag{61}$$

Additionally, we must assume values for the total mass m_1 of Earth and for its polar moment of inertia I_p. If for any range of values of r we know the corresponding values of α and β, or equivalently the values of $\phi = k/\rho_0$, then we can determine ρ_0 from Eqs. (60) and (61), provided we start from known values of density and mass at a certain level. This level is generally taken at the base of the crustal region, the so-called Moho discontinuity. This is in brief the procedure used by Jeffreys and Bullen to obtain the earliest Earth density models. As an aftermath to the Kamchatka earthquake of November 1952, emphasis has been laid upon developing techniques to improve early Earth models by using the observed frequencies of Earth vibrations triggered by earthquakes. These vibrations can be construed as being the superposition of the elastic gravitational normal modes of Earth that are excited by an earthquake. By inversion of the observed eigenperiods of Earth, one can infer an Earth model. Mathematically, the problem consists of finding density perturbations to a given Earth model so that the differences between calculated and observed eigenfrequencies can be reduced to a minimum.

The normal modes of oscillation for a nonrotating, spherically symmetric Earth are of two kinds: spheroidal and toroidal, respectively, given by

$${}_nS_l^m(\mathbf{r})\exp(i\,{}_n\omega_l t), \qquad {}_nT_l^m(\mathbf{r})\exp(i\,{}_n\varpi_l t), \tag{62}$$

where

$$\begin{aligned} {}_nS_l^m(\mathbf{r}) &= \frac{\mathbf{r}}{r}\,{}_nU_l(r)Y_l^m(\theta,\phi) + {}_nV_l(r)r\,\nabla Y_l^m(\theta,\phi). \\ {}_nT_l^m(\mathbf{r}) &= -\,{}_nW_l(r)\mathbf{r}\times\nabla Y_l^m(\theta,\phi). \end{aligned} \tag{63}$$

Here i is the imaginary unit, θ is the colatitude, ϕ is the longitude, and ∇ denotes the gradient. The Y are the normalized surface harmonics, defined as

$$\begin{aligned} Y_l^m(\theta,\phi) &= X_l^{|m|}(\theta)\exp(im\phi) \\ &= \left[\frac{(2l+1)(l-|m|)!}{4\pi(l+|m|)!}\right]^{1/2} P_l^{|m|}(\cos\theta)\exp(im\phi) \end{aligned} \tag{64}$$

in terms of the Ferrers associated Legendre functions

$$P_l^m(v) = (1 - v^2)^{m/2} \frac{d^m P_l(v)}{dv^m}, \tag{65}$$

where $v = \cos\theta$ and $|m| \leqslant l$. Notice that the absolute value of m appears in the formulation of the normalizing factor and the function of v.

These oscillations depend on three parameters: the angular order l; the azimuthal order m (with $-l \leqslant m \leqslant l$), which is related to the degree l and order m of the surface spherical harmonic $Y_l^m(\theta, \phi)$; and the radial number n, which defines the overtone.

For each value of l and n there are $2l + 1$ normal mode eigenfunctions of the spheroidal (respectively, toroidal) type all of which belong to the same eigenfrequency ${}_n\omega_l$ (respectively, ${}_n\bar{\omega}_l$): together they form a so-called multiplet. This property constitutes the main difficulty in identifying spectral peaks of the eigenfrequencies and in comparing them with the theoretical calculations. The resolution of these multiplets constitutes a major facet of the problem of ascertaining an earth model and will be briefly reviewed in what follows. Among the first models to be developed by this procedure one should acknowledge the HB1 Earth model, formulated by Haddon and Bullen in 1969. In it, 110 observations of periods of spheroidal and toroidal oscillations were taken into account based on the earthquake data of the May 1960 Chile earthquake and the one of March 1964 in Alaska. In this model, Earth's polar moment of inertia was assumed to be $I_p = 0.3309 m_1 a_1^2$, where $m_1 = 5.976 \cdot 10^{27}$ g is Earth's mass and $a_1 = 6371$ km is its mean radius, i.e., the radius of a sphere of equal volume; it was also assumed that the central density $\rho_0(0) \leqslant 13$ g/cm^3.

Another issue which has to be considered is the effect of attenuation upon the eigenperiods due to the fact that Earth is significantly anelastic over seismic frequencies. Recent research by Hart *et al.* (1977) has revealed that a frequency-dependent correction of the order of 1% should be applied to the normal mode periods to eliminate the baseline discrepancies between body wave results and normal mode results: these authors adjusted the eigenfrequencies of their earlier C2 Earth model for attenuation and subsequently by inversion obtained their QM2 model. From a more realistic point of view, however, a seismogram should be regarded as a function not only of the structure of Earth but also of the mechanism of the earthquake. In this connection, recent research by Gilbert and Dziewonski has focused on the problem of retrieving the source mechanism of an earthquake, the so-called moment rate tensor of the earthquake; this operation, coupled with the resolution of the multiplets, has been used to improve previous Earth models. These authors obtained their 1066A and B Earth models in 1975 using Earth data compiled from 1463 modes and retrieved the source mechanism of the July 1970 Colombia and August 1963 Peru–Bolivia earthquakes. Their analysis of these two deep earthquakes has revealed that the moment rate tensor possesses a significant isotropic component and that the orientation of the principal axes of the stress can change considerably during the stress release event. These facts differ considerably from the predictions of previously assumed earthquake models, such as that of a double couple; it is therefore to be expected that a procedure based upon a moment rate tensor directly determined from seismic data should be far superior to one which makes use of an a priori assumption.

We shall briefly elaborate on the problems concerning the resolution of the multiplets and retrieving the moment rate tensor of an earthquake from seismic spectra. The procedures which are currently used by seismologists are conceptually simple since they involve only linear processes; from a computational point of view they imply the inversion of symmetric matrices and lead therefore to stable numerical conditions.

From classical mechanics we know how to express the excitation of the normal modes of oscillation caused by exterior forces. In the case of an earthquake, the external force is the divergence of the moment rate tensor $M(\mathbf{r}, t)$; in the present case it will be assumed to be a point source:

$$M(\mathbf{r}, t) = M(t)\delta(\mathbf{r} - \mathbf{r}_0).$$

Let us consider first the more complex case of the spheroidal oscillations, and for typographical simplicity we shall omit, whenever possible, those indices which do not play at that time a fundamental role.

The response to the excitation can be expressed as

$$u(\mathbf{r}, t) = \sum S(\mathbf{r})\epsilon^*(\mathbf{r}_0) \cdot \mathbf{f}(t)[\exp(-\alpha t)\cos(\omega t) - 1]. \tag{66}$$

Here ω is the frequency and we have introduced an exponential attenuation term in which the attenuation parameter α is related to the quality factor Q according to $\alpha = \omega/2Q$; the asterisk denotes complex conjugacy. $\mathbf{f}(t)$ and $\epsilon(\mathbf{r})$ are two vectors whose six components coincide with the six components of the moment rate tensor M and the strain tensor E, respectively. Specifically, the correspondence of indices is meant in the following fashion:

$$\begin{aligned} &f_1 = M_{rr}, && f_2 = M_{\theta\theta}, && f_3 = M_{\phi\phi}, \\ &f_4 = M_{r\theta}, && f_5 = M_{r\phi}, && f_6 = M_{\theta\phi}, \\ &\epsilon_1 = E_{rr}, && \epsilon_2 = E_{\theta\theta}, && \epsilon_3 = E_{\phi\phi}, \\ &\epsilon_4 = 2E_{r\theta}, && \epsilon_5 = 2E_{r\phi}, && \epsilon_6 = 2E_{\theta\phi}. \end{aligned} \tag{67}$$

In these considerations, the quality factor Q is defined as

$$Q^{-1} = \frac{1}{2\pi \mathscr{E}_{\max}} \oint \frac{d\mathscr{E}}{dt}\, dt, \tag{68}$$

in terms of the energy $\mathscr{E}$ dissipated during a complete cycle of sinusoidal straining; $\mathscr{E}_{\max}$ is the peak energy. It is convenient to Fourier transform Eq. (66) and to replace the time argument with the frequency ω. We easily realize then that the response is represented by three terms: the moment rate tensor $\mathbf{f}(\omega)$, the multiplet spectrum C_k, and the transfer function A_{kj}, according to

$$u(\mathbf{r}, \omega) = \sum_k A_{kj}(\mathbf{r}) \cdot \mathbf{f}(\omega) C_k(\omega). \tag{69}$$

Here

$$A_{kj}(\mathbf{r}) = \sum_m S_k^m(\mathbf{r})[\epsilon_j^m(\mathbf{r}_0)]^* \tag{70}$$

represents the sum over the azimuthal order m since we cannot split the multiplets; j varies from 1 to 6 and k stands for the set of indices (n, l) representing a multiplet.

Before proceeding to the formulation of our two major problems, let us examine how to evaluate the components of the strain tensor E. For fixed values of the three indices (n, m, l), using the standard expressions for E as defined, e.g., by Sokolnikoff (1956), we can write the spheroidal modes as follows:

$$\begin{aligned}
\epsilon_1 &= U'Y, \\
\epsilon_2 &= \frac{1}{r}\{UY - V[(\cot\theta)Y' - m^2(\operatorname{cosec}^2\theta)Y + l(l+1)Y]\}, \\
\epsilon_3 &= \frac{1}{r}\{UY + V[(\cot\theta)Y' - m^2(\operatorname{cosec}^2\theta)Y]\}, \\
\epsilon_4 &= E_s Y', \qquad \epsilon_5 = imE_s(\operatorname{cosec}\theta)Y, \\
E_s &= (1/r)(U - V) + V', \\
\epsilon_6 &= \frac{2im}{r} V(\operatorname{cosec}\theta)(Y' - Y\cot\theta).
\end{aligned} \tag{71}$$

The primes here denote derivatives with respect to either r or θ depending on the function under consideration.

It is appropriate to use epicentral coordinates with $\theta = 0$ at the epicenter; one must then consider the limit of $\epsilon(\theta)$ as θ approaches zero. For practical purposes, we shall restrict the azimuthal order m to $|m| \leqslant \min(2, l)$, so that one has only to consider the cases $m = 0, \pm 1, \pm 2$.

Let us briefly illustrate the evaluation of the required limits. For this purpose, let us recall that $P_l^m(v)$, as defined by Eq. (65), satisfies the differential equation

$$(1 - v^2)\frac{d^2 P_l^m}{dv^2} - 2v\frac{dP_l^m}{dv} + \left[l(l+1) - \frac{m^2}{1 - v^2}\right]P_l^m = 0. \tag{72}$$

In terms of the Legendre polynomials $P_l \equiv P_l^0$, this equation becomes

$$(1 - v^2)\frac{d^{m+2}P_l}{dv^{m+2}} - 2(m+1)v\frac{d^{m+1}P_l}{dv^{m+1}} + [l(l+1) - m(m+1)]\frac{d^m P_l}{dv^m} = 0. \tag{73}$$

Knowing that $P_l(1) = 1$ and using the above equation for $m = 0$, 1, we find that

$$\begin{aligned} \left.\frac{dP_l}{dv}\right|_{v=1} &= \tfrac{1}{2}l(l+1), \\ \left.\frac{d^2 P_l}{dv^2}\right|_{v=1} &= \tfrac{1}{8}l(l+2)(l^2 - 1). \end{aligned} \tag{74}$$

Notice also that $P_l^m(1) = 0$ for $m \geqslant 1$.

For the purpose of example, let us consider the limit of ϵ_3 as θ approaches zero; we realize that we must evaluate

$$\lim_{v \to 1}\left(-v\frac{dP_l^m}{dv} - \frac{m^2}{1 - v^2}P_l^m\right) = L(m, l)$$

for $m = 0$, 1, 2. This leads to the following results:

$$\begin{aligned} L(0, l) &= -\left.\frac{dP_l}{dv}\right|_{v=1} = -\tfrac{1}{2}l(l+1), \\ L(1, l) &= \lim_{v \to 1}\left\{-\frac{d}{dv}\left[(1 - v^2)^{1/2}\frac{dP_l}{dv}\right] - (1 - v^2)^{-1/2}\frac{dP_l}{dv}\right\} \\ &= -\lim_{v \to 1}(1 - v^2)^{1/2}\frac{d^2 P_l}{dv^2} = 0, \\ L(2, l) &= \lim_{v \to 1}\left\{-\frac{d}{dv}\left[(1 - v^2)\frac{d^2 P_l}{dv^2}\right] - 4\frac{d^2 P_l}{dv^2}\right\} \\ &= -2\left.\frac{d^2 P_l}{dv^2}\right|_{v=1} = -\tfrac{1}{4}l(l+2)(l^2 - 1). \end{aligned} \tag{75}$$

The other limits can be treated in quite a similar fashion; the fundamental factor here is that both the Legendre polynomials and their derivatives remain bounded in the interval under consideration.

We introduce the set of numbers

$$b_j = \frac{1}{2^j j!}\left[\frac{(2l+1)}{4\pi}\frac{(l+j)!}{(l-j)!}\right]^{1/2} \tag{76}$$

for $0 \leqslant j \leqslant l$ and write the six components of the strain tensor as follows:

$$\begin{aligned}
m &= 0 \qquad & &\epsilon_1 = b_0 U', \qquad \epsilon_2 = \epsilon_3 = (1/r)b_0[U - \tfrac{1}{2}l(l+1)V],\\
& & &\epsilon_4 = \epsilon_5 = \epsilon_6 = 0;\\
m &= \pm 1 & &\epsilon_1 = \epsilon_2 = \epsilon_3 = \epsilon_6 = 0, \qquad \epsilon_4 = b_1 E_s,\\
& & &\epsilon_5 = imb_1 E_s;\\
m &= \pm 2 & &\epsilon_1 = \epsilon_4 = \epsilon_5 = 0, \qquad \epsilon_2 = -\epsilon_3 = (2b_2/r)V,\\
& & &\epsilon_6 = (2imb_2/r)V.
\end{aligned} \tag{77}$$

This result will ultimately allow us to evaluate the expressions $A_{kj}(\mathbf{r})$ as given by Eq. (70) for each of the six components of the strain tensor ϵ and where the sum runs for $m = -2, -1, 0, 1, 2$. For spheroidal multiplets, one can verify that

$$A_{kj}^s(\mathbf{r}) = \hat{r}U_k(r)B_j^s + \hat{\theta}V_k(r)\frac{\partial B_j^s}{\partial\theta} + \hat{\phi}\frac{V_k(r)}{\sin\theta}\frac{\partial B_j^s}{\partial\phi},$$

where

$$\begin{aligned}
B_1^s &= \epsilon_1^0 X_l^0(\theta),\\
B_2^s &= \epsilon_2^0 X_l^0(\theta) + 2\epsilon_2^2 X_l^2(\theta)\cos 2\phi,\\
B_3^s &= \epsilon_2^0 X_l^0(\theta) - 2\epsilon_2^2 X_l^2(\theta)\cos 2\phi,\\
B_4^s &= 2\epsilon_4^1 X_l^1(\theta)\cos\phi,\\
B_5^s &= 2\epsilon_4^1 X_l^1(\theta)\sin\phi,\\
B_6^s &= 4\epsilon_2^2 X_l^2(\theta)\sin 2\phi.
\end{aligned} \tag{78}$$

Here the circumflex denotes unit vectors and k stands for the set of subscripts (n, l) representative of a multiplet.

In a similar fashion, the components of the strain tensor for the toroidal modes can be written down, for fixed values of n, l, m, as

$$\begin{aligned}
\epsilon_1 &= 0, & \epsilon_2 &= (im/r)W(\operatorname{cosec}\theta)(Y' - Y\cot\theta),\\
\epsilon_3 &= -\epsilon_2, & \epsilon_4 &= imE_t Y\operatorname{cosec}\theta, \qquad E_t = W' - (W/r),\\
\epsilon_5 &= -Y'E_t, & \epsilon_6 &= (W/r)[2(\cot\theta)Y' - 2m^2(\operatorname{cosec}^2\theta)Y + l(l+1)Y].
\end{aligned} \tag{79}$$

At the epicenter $\theta = 0$, this will reduce to the following:

$$\begin{aligned}
&m = 0 \qquad && \epsilon_1 = \epsilon_2 = \epsilon_3 = \epsilon_4 = \epsilon_5 = \epsilon_6 = 0; \\
&m = \pm 1 \qquad && \epsilon_1 = \epsilon_2 = \epsilon_3 = \epsilon_6 = 0, \qquad \epsilon_4 = imb_1 E_t, \\
& && \epsilon_5 = -b_1 E_t; \\
&m = \pm 2 \qquad && \epsilon_1 = \epsilon_4 = \epsilon_5 = 0, \qquad \epsilon_2 = imb_2 W/r, \\
& && \epsilon_3 = -\epsilon_2, \qquad \epsilon_6 = -4b_2 W/r.
\end{aligned} \tag{80}$$

For toroidal multiplets, we have then the representation

$$A^t_{kj}(\mathbf{r}) = \frac{\hat{\theta}}{\sin\theta} W(r) \frac{\partial B^t_j}{\partial \phi} - \hat{\phi} W(r) \frac{\partial B^t_j}{\partial \theta},$$

where

$$\begin{aligned}
&B^t_1 = 0, && B^t_4 = -2\epsilon^1_5 X^1_l(\theta)\sin\phi, \\
&B^t_2 = -\epsilon^2_6 X^2_l(\theta)\sin 2\phi, && B^t_5 = 2\epsilon^1_5 X^1_l(\theta)\cos\phi, \\
&B^t_3 = \epsilon^2_6 X^2_l(\theta)\sin 2\phi, && B^t_6 = 2\epsilon^2_6 X^2_l(\theta)\cos 2\phi.
\end{aligned} \tag{81}$$

Let us now approach the question of how to resolve a multiplet. The fundamental relation which expresses the response of the normal modes to an excitation can be written as follows:

$$u(\mathbf{r}, \omega) = \sum_k A_k(\mathbf{r}, \omega) C_k(\omega), \tag{82}$$

where

$$A_k(\mathbf{r}, \omega) = A_{kj}(\mathbf{r}) \cdot \mathbf{f}_j(\omega) = \sum_{j=1}^{6} A_{kj}(\mathbf{r}) f_j(\omega) \tag{83}$$

represents the geometrical shape of the kth multiplet due to the excitation f. Superscript s or t shall be used to distinguish between spheroidal and toroidal multiplets. Let us make the following assumptions: (1) we can analyze a large number of seismograms—this means that the functions $u(\mathbf{r}, \omega)$ are available; (2) we know an approximate Earth model (which, of course, we intend to improve)—this means that we know the set $S(\mathbf{r})$, $T(\mathbf{r})$ and can compute therefrom the A_{kj} vectors; and (3) we also have available an approximate model of the source mechanism, that is, $f_j(\omega)$. If so, we evaluate $A_k(\mathbf{r}, \omega)$ and can solve for $C_k(\omega)$ from Eq. (82). For this purpose, we proceed as follows.

We introduce the quantities

$$
\begin{aligned}
V_k(\omega) &= \int A_k^*(\mathbf{r},\omega)\cdot u(\mathbf{r},\omega)\,d\Omega \\
&= \sum_h \left[\int A_k^*(\mathbf{r},\omega)\cdot A_h(\mathbf{r},\omega)\,d\Omega\right] C_h(\omega) \\
&= \sum_h B_{kh}(\omega) C_h(\omega),
\end{aligned}
\tag{84}
$$

where the integration is over the whole unit sphere; both k and h vary over the set of values assumed by (n, l). The matrix B is a block diagonal matrix because of the orthogonality conditions valid among certain types of spherical harmonics. We shall invert B and write symbolically

$$C(\omega) = B^{-1}V(\omega), \tag{85}$$

where the right-hand side is a known matrix whose frequency has been computed from the given Earth model, source mechanism, and seismograms. This will allow us to identify multiplets and determine their peak eigenfrequencies ω_k. This process is known in seismological parlance as stripping; it is essentially a phase equalization procedure whereby we recover the spectrum of the multiplets in terms of the spectrum we have evaluated based on the Earth model we have used and on the readings of seismograms recorded by a global network of stations. In the practical application of the above procedure, several factors must be taken into consideration which could play an important role in vitiating the quality of the results. Among these, we must consider the followings: adverse effects due to the limited dynamic range of the recording systems; the geographical distribution of the network of recording stations (incomplete geographical coverage might upset the block diagonal character of the B-matrix); the form of the recordings, which are analog and must be digitized, so that we cannot ignore the effect of starting and stopping digitization; signal loss, which causes recordings to be truncated; the error caused by the use of an approximate Earth model and source mechanism; the necessity of assuming a Q model; and the width of the frequency band to be stripped for each multiplet, which must be chosen according to criteria having overall significance for the operation we are envisaging. Let us proceed now with some remarks concerning the retrieval of the moment rate tensor of an earthquake starting from the assumptions that we (1) know the peak frequency of a large number of multiplets, (2) have an approximate Earth model, and (3) are able to analyze numerous seismograms. For this purpose, let us rewrite the response as follows:

$$u(\mathbf{r},\omega) = \sum_k A_{kj}(\mathbf{r})\cdot \mathbf{f}(\omega) C_k(\omega), \tag{86}$$

where k, as usual, refers to the set of indices (n, l) representing a multiplet and j varies from 1 to 6. We know all the elements appearing in Eq. (86) and must retrieve the vector $\mathbf{f}$ having the six components $f_j(\omega)$.

Let us introduce the following algorithm, which goes by the name of stacking:

$$V_{kj}(\omega) = \int A_{kj}^{\mathrm{T}}(\mathbf{r}) \cdot u(\mathbf{r}, \omega)\, d\Omega.$$

Here the integration is over the whole unit sphere and the superscript T stands for the operation of matrix transposition. Substitution from Eq. (86) gives

$$V_{kj}(\omega) = \sum_h \left[\int A_{kj}^{\mathrm{T}}(\mathbf{r}) \cdot A_{hr}(\mathbf{r})\, d\Omega \right] \cdot \mathbf{f}(\omega) C_h(\omega),$$

where h extends to the sum of multiplets and r varies from 1 to 6. For fixed values of k and h, the expression within brackets is a 3×3 symmetric matrix S_{khjr} which depends on the spherical harmonics $Y_k Y_h$; because of the orthogonality of these functions, this matrix is predominantly diagonal. For a fixed multiplet k, of a given type (i.e., spheroidal or toroidal), the summation over h must reduce to terms for which $h = k$ and apply to the same type of multiplet. Thus for a fixed multiplet k, we reach the vector

$$V_{kj}(\omega) = \sum_{r=1}^{6} S_{kkjr} f_r(\omega) C_k(\omega) \tag{87}$$

having six components: $j = 1, 2, \ldots, 6$.

Let us now apply to the previous equation the integral operator

$$I(f) = \int_{\omega_k - w}^{\omega_k + w} f(\omega)\, d\omega,$$

where ω_k is the peak frequency of the kth multiplet and w is taken to be much larger than the dissipation coefficient α_k of the multiplet. Since $C_k(\omega)$ is obtained as the Fourier transform of

$$\exp(-\alpha_k t) \cos(\omega_k t) - 1,$$

we know that

$$I[C_k(\omega)] = \arctan(w/\alpha_k)$$

and that for large values of w this approaches the limit $\pi/2$. Since the multiplet has a narrow spread around the peak ω_k, the integral of C_k can be evaluated and will not vary much with w; this means that we can in effect remove the contribution due to dissipation. The value of the integral, denoted by β_k, can be safely taken in the neighborhood of 1.25.

Applying the integral operator I to Eq. (87), we find

$$I[V_{kj}(\omega)] = \sum_{r=1}^{6} \beta_k S_{kkjr} f_r(\omega). \tag{88}$$

Given a frequency band, we assume that $\mathbf{f}(\omega)$ is constant within that band, and we sum the previous equation (88) over the set of multiplets encompassed within such a band:

$$\sum_k \beta_k^{-1} I[V_{kj}(\omega)] = \sum_{r=1}^{6} \sum_k S_{kkjr} f_r(\omega).$$

This can be written as

$$h_j = \sum_{r=1}^{6} S_{jr} f_r(\omega), \tag{89}$$

with the obvious interpretation of the notation. We get a system of six linear equations in the six components $f_1, \ldots, f_6$ of the moment rate tensor valid within that frequency band.

In applying these considerations to real data, one is of course faced with the same type of difficulties and shortcomings mentioned when we treated the resolution of multiplets.

We now provide some detailed information on the iterative procedure whereby older Earth models evolved into the 1066A and B models obtained by Gilbert and Dziewonski in 1975. These two models are based on the analysis of two deep earthquakes: the Colombia earthquake of July 31, 1970, which had a depth of 651 km and a magnitude of 7.1, and the Peru–Bolivia earthquake of August 15, 1963, which had a depth of 543 km and an estimated magnitude between 6 and 7.9. The network used for the analysis of these events was the World-Wide Standard Seismograph Network, which provided 165 recordings from 65 stations for the 1970 event and 46 recordings from 37 stations for the 1963 event.

All seismograms recorded in the period from the onset of the event until 2 days later were examined and a number of them chosen for digitization; the amplitude spectra of the Fourier transforms of the digitized records were examined, and records showing poor signal-to-noise ratios were rejected. The Fourier spectra of the chosen recordings were then corrected for instrumental response and phase shift resulting from delay in the beginning of digitization with respect to the origin of time.

An existing Earth model, the C196 based on an Alaska earthquake, was employed as a starting model, and the preliminary estimates of those two deep earthquakes, obtained by Mendiguren (1973) and Chandra (1970), were also used. The previously exposed procedure of stripping provided the means of identifying a set of 506 normal modes with periods from 83 to 540 sec and overtones n ranging from 0 to 4 for toroidal modes and from 0 to 30 for spheroidal modes. This set, together with the value of Earth's mass and polar moment of inertia, gave rise to the so-called Earth model 508 of 1973.

Next, the spectra of both earthquakes and the new Earth model 508 were used to improve the estimates of their moment rate tensors. For this operation, 1348 multiplets were used with angular order l up to 30 for toroidal and 40 for spheroidal modes.

The following remarks support the conclusions that were reached.

1. Had the moment rate tensor been a delta function of time, its Fourier cosine transform would have been a constant and its Fourier sine transform would have been zero. The actual behavior of this tensor was found to be quite different; hence it was rightly inferred that the stress release occurred over a finite period of time, which was estimated to have been of the order of tens of seconds.

2. The standard errors for the off-diagonal components of the tensor (i.e., f_4, f_5, f_6) were found to be about three times less than those pertaining to the diagonal elements (f_1, f_2, f_3). This suggested that the excitation pattern of these events had been more sensitive to the shear elements than the diagonal elements. It was therefore deemed appropriate to approximate the observed excitation by means of a tensor whose components constitute a matrix with zero trace. However, the isotropic part of the tensor was found to be statistically significant; e.g., for the Colombia earthquake and $\omega = 0.015 \text{ sec}^{-1}$ it was found that

$$\tfrac{1}{3}\operatorname{Re}(f_1 + f_2 + f_3) = -6.0 \cdot 10^{27} \qquad \text{dyn cm},$$

with a standard error about 10 times smaller, which precluded any possibility of it being a chance event. The decomposition of the tensor $M(t)$ was thus considered to be of the type

$$M(t) = Im(t) + D(t),$$

where I is the 3×3 identity matrix, $m(t) = \frac{1}{3}\operatorname{Re}(f_1 + f_2 + f_3)$, and $D(t)$ is a deviator, i.e., a matrix with zero trace.

3. It was next inferred that the isotropic stress $m(t)$ was compressive and had preceded the onset of deviatoric stress by over 80 sec. In fact, $m(\omega)$ was found to be an even function of ω; this means that $m(t)$ is also an even function of time, so that $m(t)$ has to be precursive. The inference that $m(\omega)$ is an even function was based essentially on the smoothness of the data at low frequencies. Values of the moment rate tensors were not available, however,

Table 2.3

Values of Density for the 1066A Earth Model

Depth (km)	ρ (g/cm^3)	Depth	ρ	Depth	ρ	Depth	ρ
0	2.183	1,121	4.635	2,576	5.421	5,142	12.153
6	2.183	1,156	4.660	2,610	5.434	5,142	13.021
11	2.183	1,191	4.683	2,645	5.447	5,181	13.031
11	3.343	1,225	4.705	2,680	5.460	5,219	13.045
37	3.351	1,260	4.724	2,714	5.471	5,257	13.060
62	3.358	1,294	4.741	2,749	5.483	5,296	13.074
88	3.365	1,329	4.756	2,784	5.495	5,334	13.088
114	3.372	1,364	4.770	2,818	5.506	5,373	13.102
139	3.379	1,398	4.783	2,853	5.518	5,411	13.116
165	3.387	1,433	4.797	2,887	5.528	5,449	13.131
190	3.393	1,468	4.811	2,887	9.914	5.488	13.145
216	3.402	1,502	4.825	2,958	10.028	5.526	13.160
242	3.419	1,537	4.841	3,028	10.134	5,565	13.175
267	3.443	1,571	4.857	3,099	10.235	5,603	13.190
293	3.473	1,606	4.874	3,169	10.333	5,641	13.205
319	3.509	1,641	4.892	3,240	10.427	5,680	13.220
344	3.551	1,675	4.911	3,310	10.516	5,718	13.235
370	3.600	1,710	4.930	3,381	10.603	5,757	13.251
395	3.657	1,745	4.949	3,451	10.687	5,795	13.267
421	3.712	1,779	4.969	3,522	10.772	5,833	13.284
421	3.712	1,814	4.988	3,592	10.858	5,872	13.300
452	3.764	1,849	5.008	3,663	10.946	5,910	13.316
484	3.805	1,883	5.027	3,733	11.033	5,949	13.331
515	3.850	1,918	5.046	3,803	11.116	5,987	13.345
546	3.903	1,952	5.066	3,874	11.192	6,025	13.357
577	3.961	1,987	5.086	3,944	11.265	6,064	13.368
609	4.025	2,022	5.106	4,015	11.335	6,102	13.377
640	4.106	2,056	5.127	4,085	11.406	6,141	13.385
671	4.208	2,091	5.147	4,156	11.475	6,179	13.393
671	4.208	2,126	5.169	4,226	11.541	6,217	13.399
706	4.319	2,160	5.190	4,297	11.602	6,256	13.406
740	4.403	2,195	5.212	4,367	11.660	6,294	13.411
775	4.468	2,229	5.233	4,438	11.716	6,333	13.418
810	4.511	2,264	5.255	4,508	11.769	6,371	13.421
844	4.535	2,299	5.277	4,578	11.818		
879	4.537	2,333	5.298	4,649	11.863		
913	4.537	2,368	5.318	4,719	11.904		
948	4.542	2,403	5.338	4,790	11.944		
983	4.552	2,437	5.357	4,860	11.985		
1,017	4.567	2,472	5.374	4,931	12.026		
1,052	4.587	2,507	5.391	5,001	12.068		
1,087	4.610	2,541	5.406	5,072	12.110		

for frequencies $|\omega| \leqslant 0.0125\ \mathrm{sec}^{-1}$; the smoothness of the data in such a range can nevertheless be established by using a property that holds for a pair $f(t)$ and $f(\omega)$ of Fourier transforms:

$$\lim_{\omega \to 0} \omega f(\omega) = \lim_{t \to \infty} f(t).$$

Oscillations of $f(\omega)$ for small values of ω should give rise to oscillations of $f(t)$ for long time periods. Since the stress release of these two events was primarily concentrated near $t = 0$, one can safely rule out fluctuations for long time periods, which accordingly ensures smoothness of $f(\omega)$ for low frequencies.

4. Finally, it was ascertained that for both events the principal axes of $M(t)$ did change their angular positions in space during the events, creating a situation quite different from the usual earthquake models used in previous computations.

Next, by using the 165 spectra of the Colombia earthquake, the 46 spectra of the Peru–Bolivia earthquake, the newly obtained expressions for the source mechanisms of these two earthquakes, and the 508 Earth model, a set of 812 eigenfrequencies were identified.

The final operation consisted of using models 508 and B1 of Jordan and Anderson and a set of 1066 distinct eigenfrequencies to arrive at models 1066A and B. The difference between the two models is primarily the smoothness exhibited by model A in the mantle, which is not the case for model B. Travel time data were not used to derive either model A or B.

Tables 2.3–2.5 provide only density data for the 1066A, HB1, and QM2 models; they were used in the numerical integration of the Clairaut equation, which will be discussed in a forthcoming section of this chapter.

In closing this section we want to remark that the major discontinuities encountered in the existing Earth models are located where classical ray

Table 2.4

Values of Density for the HB1 Earth Model

Depth (km)	ρ (g/cm^3)	Depth	ρ	Depth	ρ
0	2.840	984	4.529	3600	10.948
15	2.840	1000	4.538	3800	11.176
15	3.313	1200	4.655	4000	11.383
60	3.332	1400	4.768	4200	11.570
100	3.348	1600	4.877	4400	11.737
200	3.387	1800	4.983	4600	11.887
300	3.424	2000	5.087	4800	12.017
350	3.441	2200	5.188	4982	12.121
350	3.700	2400	5.288	5000	12.130
400	3.775	2600	5.387	5121	12.197
413	3.795	2800	5.487	5200	12.229
500	3.925	2878	5.527	5400	12.301
600	4.075	2878	9.927	5600	12.360
650	4.150	3000	10.121	5800	12.405
650	4.200	3200	10.421	6000	12.437
800	4.380	3400	10.697	6200	12.455
				6371	12.460

Table 2.5

Values of Density for the QM2 Earth Model

Depth (km)	ρ (g/cm^3)	Depth	ρ	Depth	ρ	Depth	ρ
0	1.02	496	3.72	1546	4.85	3471	10.78
3	1.02	521	3.73	1621	4.89	3571	10.91
3	2.80	546	3.89	1696	4.94	3671	11.02
21	2.80	571	3.95	1771	4.97	3771	11.12
21	3.49	596	3.97	1846	5.01	3871	11.21
41	3.50	621	3.99	1921	5.05	3971	11.29
61	3.52	646	4.00	1996	5.09	4071	11.37
81	3.45	671	4.04	2071	5.14	4171	11.45
101	3.39	671	4.38	2146	5.19	4271	11.53
121	3.31	696	4.40	2221	5.24	4471	11.69
146	3.29	711	4.42	2296	5.29	4571	11.78
171	3.31	728	4.43	2371	5.34	4671	11.85
196	3.33	746	4.44	2446	5.38	4771	11.93
221	3.35	769	4.47	2521	5.42	4871	11.99
246	3.36	798	4.51	2596	5.45	4971	12.05
271	3.36	821	4.52	2671	5.48	5071	12.09
296	3.38	871	4.55	2746	5.49	5156	12.12
321	3.43	946	4.58	2821	5.51	5156	12.30
346	3.51	1021	4.61	2861	5.52	5371	12.48
371	3.59	1096	4.64	2886	5.52	5571	12.52
388	3.63	1171	4.68	2886	9.97	5771	12.52
404	3.71	1246	4.71	2971	10.10	5971	12.52
421	3.82	1321	4.74	3071	10.24	6071	12.53
446	3.81	1396	4.77	3171	10.38	6271	12.57
471	3.76	1471	4.81	3371	10.65	6371	12.57

theory and normal mode theory diverge. A more sophisticated mathematical theory of wave propagation should be developed to resolve those propagation phenomena that present-day wave theory seems inadequate to handle.

2.06 INTRODUCTION TO NUMERICAL METHODS

We plan to discuss briefly certain numerical operators required for the numerical integration of ordinary differential equations. More specifically, we shall consider the central difference operator δ and shall express both the first and second derivatives of a function at a pivotal point of an integration grid in terms of powers of δ.

We start by defining the incremental operator E in terms of the differential operator D by using a Taylor series expansion:

$$Ey(x) = y(x+h) = y(x) + hy'(x) + \cdots + \frac{h^n}{n!} y^{(n)}(x) + \cdots$$
$$= \exp(hD)\, y(x).$$

Here the nth power of D applied to $y(x)$ should be construed as the nth derivative of $y(x)$. Thus we get

$$E = \exp(hD). \tag{90}$$

Powers of E should be interpreted in terms of multiple incrementation:

$$E^p y(x) = y(x + ph).$$

In particular, the central operator δ yields

$$\delta y(x) = y(x + \tfrac{1}{2}h) - y(x - \tfrac{1}{2}h) = (E^{1/2} - E^{-1/2})y(x). \tag{91}$$

We also define the averaging operator as

$$\mu y(x) = \tfrac{1}{2}[y(x + \tfrac{1}{2}h) + y(x - \tfrac{1}{2}h)] = \tfrac{1}{2}(E^{1/2} + E^{-1/2})y(x). \tag{92}$$

It is a well-known fact that such operators can be handled according to the elementary rules of algebra and calculus. Accordingly, using Eqs. (91), (92), and (90), we can write

$$\begin{aligned} \delta &= E^{1/2} - E^{-1/2} = \exp(\tfrac{1}{2}hD) - \exp(-\tfrac{1}{2}hD) \\ &= 2\sinh(\tfrac{1}{2}hD), \\ \mu &= \tfrac{1}{2}(E^{1/2} + E^{-1/2}) = \cosh(\tfrac{1}{2}hD). \end{aligned} \tag{93}$$

Since $\cosh^2 x = 1 + \sinh^2 x$,

$$\mu^2 = 1 + \tfrac{1}{4}\delta^2. \tag{94}$$

It is customary to use the following notation for the pivotal points of the grid: a subscript 0 for a given point x; a subscript 1 for the point $x + h$; and a subscript -1 for the point $x - h$. The same considerations apply to the values of a function at those points; thus

$$f(x) = f_0, \qquad f(x + h) = f_1, \qquad f(x - h) = f_{-1}, \qquad y'(x) = y'_0,$$

and so on.

Let us remark at this stage that in dealing with central difference formulas, if one wants to operate within the elements of a grid and avoid introducing halfway points, one should limit oneself to either even powers of δ or products of the operator μ and odd powers of δ. Thus, e.g.

$$\delta^2 y_0 = y_1 - 2y_0 + y_{-1}, \qquad \mu\delta y_0 = \tfrac{1}{2}(y_1 - y_{-1}). \tag{95}$$

We next use Eq. (93) to express the operator D as an infinite series of the central difference operator δ:

$$hD = 2\sinh^{-1}(\tfrac{1}{2}\delta) = \delta - \tfrac{1}{24}\delta^3 + \tfrac{3}{640}\delta^5 - \cdots.$$

To remain on a preassigned grid, we shall multiply its right-hand side by the operator $\mu(1 + \frac{1}{4}\delta^2)^{-1/2}$, which, according to Eq. (94), is the identity operator, where

$$(1 + \tfrac{1}{4}\delta^2)^{-1/2} = 1 - \tfrac{1}{8}\delta^2 + \tfrac{3}{128}\delta^4 - \cdots.$$

We thus obtain the useful formula

$$hy'_0 = \mu\delta y_0 - \tfrac{1}{6}\mu\delta^3 y_0 + \tfrac{1}{30}\mu\delta^5 y_0 - \cdots. \tag{96}$$

Similarly,

$$h^2D^2 = [2\sinh^{-1}(\tfrac{1}{2}\delta)]^2,$$

which gives

$$h^2 y''_0 = \delta^2 y_0 - \tfrac{1}{12}\delta^4 y_0 + \tfrac{1}{90}\delta^6 y_0 - \cdots. \tag{97}$$

By using the preceding formulas, we can represent an ordinary differential equation of the second order by means of a system of finite difference equations, one for each point of the integration grid. It will be a simultaneous system since even if we truncate formulas by dropping terms in δ^3, we are introducing the values of the function at the two neighboring points: y_{-1}, y_0, y_{+1}. We consider the second-order differential equation

$$y'' + f(x)y' + g(x)y = k(x)$$

and use the notation introduced earlier. We can easily show that at a pivotal point x the following relation will hold:

$$(1 - \tfrac{1}{2}hf_0)y_{-1} - (2 - h^2 g_0)y_0 + (1 + \tfrac{1}{2}hf_0)y_1 = h^2 k_0 - Cy_0. \tag{98}$$

Here h is the width of the grid and C stands for the sum of the third- and higher-order powers of the central difference operator:

$$C = (-\tfrac{1}{12}\delta^4 + \tfrac{1}{90}\delta^6 - \cdots) + hf_0(-\tfrac{1}{6}\mu\delta^3 + \tfrac{1}{30}\mu\delta^5 - \cdots).$$

2.07 NUMERICAL SOLUTION OF THE CLAIRAUT EQUATION FOR VARIOUS EARTH MODELS

The first integration of the Clairaut equation using a realistic Earth density profile was undertaken by James and Kopal in 1963; they used the 1940 Bullen Earth model. The most comprehensive numerical results on the subject are due to Lanzano and Daley (1977); they used the HB1 model of Haddon and Bullen, the 1066A model, and the QM2 model. In what follows, we shall briefly describe some details of the numerical integration procedure followed by these authors, which depends on the central difference operator discussed in the previous section.

The six unknown deformation coefficients $f_{21}, f_{22}, f_{23}, f_{42}, f_{43}$, and f_{63} defining hydrostatic equilibrium up to third-order terms satisfy the Clairaut equation for various orders of approximation:

$$a^2 f'' + 6aDf' + (6D - C)f = R(a) \tag{99}$$

and the boundary conditions

$$f(0) = 0, \qquad Af(a_1) + Bf'(a_1) = S(a_1) \tag{100}$$

at both ends of the integration interval, which extends from $a = 0$ to $a_1 =$ 6371 km. Here the primes denote derivatives with respect to the mean radius a; A, B, C are constants; and $R(a)$, $S(a)$ are functions of a, both sets depending on lower order approximations, whose expressions have been explicitly given by Eqs. (19) and (20) and Tables 2.1 and 2.2. $D(a)$ is the ratio of ρ to $\bar{\rho}$ and can be shown to approach unity as the radius tends to zero, i.e., $D(0) = 1$.

Because each of the density models presents discontinuities, one has to evaluate $\bar{\rho}(a)$, which is defined as an integral by piecewise integration. This requires breaking the integration interval at each point c of discontinuity, thus taking the integral from zero to c and from c to the generic point a, provided that one then evaluates the limit both from the left and from the right of the point c. $\bar{\rho}(a)$ can then be evaluated along a certain grid, and the values at intermediate points can be filled by Lagrangian interpolation between the data points.

Using the central difference formulas discussed in the previous section [see Eq. (98)] and choosing an interval h, we can write the finite difference form of the Clairaut equation at a pivotal point a_j^* as

$$\left(1 - \frac{3h}{a_j^*} D_j\right) f_{j-1} + \left[-2 + \left(\frac{h}{a_j^*}\right)^2 (6D_j - C)\right] f_j + \left(1 + \frac{3h}{a_j^*} D_j\right) f_{j+1} = \left(\frac{h}{a_j^*}\right)^2 R_j, \tag{101}$$

where third- and higher-order central difference terms have been neglected. We select N pivotal points a_j^* $(j = 1, 2, \ldots, N)$ at equal intervals such that $a_1^* = h$ and $a_N^* = a_1 = 6371$ km. This choice requires the value f_{N+1} of f at the external point $a = a_1 + h$. This extraneous quantity can be eliminated, however, by making use of the boundary condition valid at the surface, which contains the derivative f'. In fact, using Eqs. (96) and (95), we can see that in the first approximation

$$f'_j = \frac{1}{2h}(f_{j+1} - f_{j-1}).$$

Thus the boundary condition at $a = a_N^*$ can be written as

$$Af_N + \frac{B}{2h}(f_{N+1} - f_{N-1}) = S_N,$$

and this allows us to solve for f_{N+1}.

For the remaining pivotal points ($j = 1, 2, \ldots, N - 1$), since $a_j^* = jh$, we can write the system of linear equations

$$\left(1 - \frac{3}{j} D_j\right) f_{j-1} + \left[-2 + \frac{1}{j^2}(6D_j - C)\right] f_j + \left(1 + \frac{3}{j} D_j\right) f_{j+1} = \frac{R_j}{j^2}. \tag{102}$$

In particular, for $j = 1$, since $f_0 = f(0) = 0$,

$$(6D_1 - C - 2)f_1 + (1 + 3D_1)f_2 = R_1.$$

Table 2.6

Deformation Coefficients for the Equipotential Surfaces According to the HB1 Earth Model: Solution of the Clairaut Equations

Depth (km)	$-10^6 \cdot f_0$	$-10^2 \cdot f_2$	$+10^5 \cdot f_4$	$-10^8 \cdot f_6$
0	0.99421	0.22298	0.44766	0.96569
15	0.99147	0.22268	0.44630	0.96106
300	0.94109	0.21695	0.42139	0.87871
350	0.93259	0.21596	0.41724	0.86551
650	0.88288	0.21013	0.39312	0.79141
900	0.84214	0.20522	0.37324	0.73257
1200	0.79332	0.19918	0.34890	0.66167
1500	0.74495	0.19302	0.32400	0.58968
1800	0.69804	0.18684	0.29891	0.51695
2100	0.65431	0.18089	0.27448	0.44533
2400	0.61654	0.17559	0.25242	0.37940
2878	0.58072	0.17041	0.23085	0.31418
3000	0.57747	0.16994	0.22905	0.30954
3300	0.57030	0.16888	0.22508	0.29935
3600	0.56416	0.16797	0.22167	0.29071
3900	0.55893	0.16719	0.21876	0.28338
4200	0.55451	0.16652	0.21628	0.27720
4500	0.55084	0.16597	0.21422	0.27205
4800	0.54786	0.16552	0.21253	0.26785
5121	0.54544	0.16516	0.21114	0.26440
5400	0.54385	0.16492	0.21023	0.26215
5700	0.54260	0.16473	0.20952	0.26038
6000	0.54182	0.16461	0.20907	0.25927
6300	0.54149	0.16456	0.20888	0.25880
6371	0.54147	0.16455	0.20887	0.25878

Table 2.7

Deformation Coefficients for the Equipotential Surfaces According to the 1066A Earth Model: Solution of the Clairaut Equations

Depth (km)	$-10^6 \cdot f_0$	$-10^2 \cdot f_2$	$+10^5 \cdot f_4$	$-10^8 \cdot f_6$
0	0.99376	0.22293	0.44762	0.96520
11	0.99175	0.22271	0.44662	0.96181
300	0.94067	0.21690	0.42135	0.87799
421	0.92023	0.21453	0.41137	0.84631
671	0.87929	0.20970	0.39158	0.78609
900	0.84230	0.20524	0.37364	0.73341
1200	0.79357	0.19922	0.34946	0.66318
1500	0.74509	0.19303	0.32462	0.59144
1800	0.69795	0.18683	0.29953	0.51888
2100	0.65390	0.18084	0.27511	0.44758
2400	0.61557	0.17545	0.25295	0.38177
2887	0.57738	0.16992	0.23032	0.31392
3000	0.57402	0.16943	0.22856	0.30950
3300	0.56562	0.16818	0.22422	0.29879
3600	0.55776	0.16701	0.22026	0.28929
3900	0.55019	0.16587	0.21661	0.28088
4200	0.54243	0.16470	0.21302	0.27310
4500	0.53381	0.16339	0.20913	0.26512
4800	0.52371	0.16183	0.20435	0.25515
5142	0.51382	0.16030	0.19902	0.24213
5400	0.51187	0.15999	0.19801	0.23985
5700	0.50964	0.15964	0.19688	0.23738
6000	0.50787	0.15937	0.19600	0.23542
6300	0.50686	0.15921	0.19552	0.23442
6371	0.50674	0.15919	0.19547	0.23430

The integration interval was subdivided into subintervals by the points of discontinuity; these points of the model were assumed as fixed pivotal points and were subsequently included in any finer grid required for convergence purposes.

Because the third- and higher-order central differences were neglected, an equation of variation had to be solved to compute a correction to the various f's.

The computations were programmed in Fortran and executed on the Texas Instrument computer at NRL in double-precision floating point arithmetic to a precision of 16 decimal digits. By iteration, results were obtained that converged to at least 5 decimal digits.

Tables 2.6, 2.7, and 2.8 provide the deformation coefficients f_0, f_2, f_4, and f_6 of the equipotential surfaces [Eq. (6)] for the HB1, 1066A and QM2, density models, respectively. The numbers N of pivotal points required to reach the mentioned accuracy were 851, 580, and 607, respectively. The

Table 2.8

Deformation Coefficients for the Equipotential Surfaces According to the QM2 Earth Model: Solution of the Clairaut Equations

Depth (km)	$-10^6 \cdot f_0$	$-10^2 \cdot f_2$	$+10^5 \cdot f_4$	$-10^8 \cdot f_6$
0	0.99239	0.22278	0.44648	0.96148
3	0.99184	0.22272	0.44620	0.96055
21	0.98857	0.22235	0.44457	0.95501
300	0.93931	0.21674	0.42010	0.87363
671	0.87834	0.20959	0.39047	0.78217
900	0.84170	0.20517	0.37269	0.73006
1200	0.79336	0.19919	0.34864	0.66024
1500	0.74531	0.19306	0.32392	0.58877
1800	0.69871	0.18693	0.29903	0.51666
2100	0.65530	0.18103	0.27485	0.44588
2400	0.61776	0.17577	0.25301	0.38077
2886	0.58122	0.17049	0.23104	0.31431
3000	0.57827	0.17005	0.22942	0.31013
3300	0.57118	0.16901	0.22555	0.30029
3600	0.56497	0.16809	0.22222	0.29198
3900	0.55927	0.16724	0.21921	0.28471
4200	0.55385	0.16643	0.21640	0.27801
4500	0.54875	0.16566	0.21377	0.27189
4800	0.54396	0.16493	0.21129	0.26620
5156	0.53932	0.16423	0.20866	0.25960
5400	0.53853	0.16411	0.20807	0.25791
5700	0.53832	0.16407	0.20783	0.25716
6000	0.53770	0.16398	0.20751	0.25642
6300	0.53681	0.16384	0.20707	0.25546
6371	0.53681	0.16384	0.20707	0.25545

Table 2.9

Comparison of Various Earth Density Models[a]

Parameter	QM2	1066A	HB_1	Bullen (1940)
$f_0\ (a_1)$	$-0.99239 \cdot 10^{-6}$	$-0.99376 \cdot 10^{-6}$	$-0.99421 \cdot 10^{-6}$	$-0.10098 \cdot 10^{-5}$
$f_2\ (a_1)$	$-0.22278 \cdot 10^{-2}$	$-0.22293 \cdot 10^{-2}$	$-0.22298 \cdot 10^{-2}$	$-0.22473 \cdot 10^{-2}$
$f_4\ (a_1)$	$+0.44648 \cdot 10^{-5}$	$+0.44762 \cdot 10^{-5}$	$+0.44766 \cdot 10^{-5}$	$+0.45202 \cdot 10^{-5}$
$f_6\ (a_1)$	$-0.96148 \cdot 10^{-8}$	$-0.96520 \cdot 10^{-8}$	$-0.96569 \cdot 10^{-8}$	$-0.97129 \cdot 10^{-8}$
$K_2\ (a_1)$	$-0.10735 \cdot 10^{-2}$	$-0.10751 \cdot 10^{-2}$	$-0.10756 \cdot 10^{-2}$	$-0.10929 \cdot 10^{-2}$
$K_4\ (a_1)$	$+0.29625 \cdot 10^{-5}$	$+0.29763 \cdot 10^{-5}$	$+0.29776 \cdot 10^{-5}$	$+0.30495 \cdot 10^{-5}$
$K_6\ (a_1)$	$-0.11199 \cdot 10^{-7}$	$-0.11284 \cdot 10^{-7}$	$-0.11293 \cdot 10^{-7}$	$-0.11598 \cdot 10^{-7}$
ϵ^{-1}	299.8	299.6	299.6	297.2
$A = Ie$	$0.80087 \cdot 10^{35}$	$0.80168 \cdot 10^{35}$	$0.80212 \cdot 10^{35}$	$0.80893 \cdot 10^{35}$
$C = Ip$	$0.80347 \cdot 10^{35}$	$0.80429 \cdot 10^{35}$	$0.80473 \cdot 10^{35}$	$0.81159 \cdot 10^{35}$
$(C-A)/C$	$0.3239 \cdot 10^{-2}$	$0.3242 \cdot 10^{-2}$	$0.3243 \cdot 10^{-2}$	$0.3274 \cdot 10^{-2}$

[a] $q = 0.00115$.

results tabulated at the chosen values of the depth are linear interpolations between the pivotal points.

Table 2.9 shows the surface values of the f and of the coefficients K appearing in the spherical harmonic expansion of the exterior potential according to Eq. (29). Also shown are the values of the ellipticity, defined as

$$\epsilon = \frac{r(a_1, 0) - r(a_1, 1)}{r(a_1, 0)}, \tag{103}$$

and the ratio $(C - A)/C$, where A is the moment of inertia with respect to any barycentric axis in the equatorial plane and C is the moment of inertia with respect to the polar axis. This table is a comparison of four Earth models and includes the results for the 1940 Bullen model, which required 387 pivotal points to reach the same accuracy. In all these computations, it was assumed that $q = 0.00115$ is the value of the rotational parameter; the moments of inertia are expressed in grams per kilometer squared. It is interesting to remark that the geodetic values of these parameters obtained from satellite measurements lie between the data for the Bullen and HB1 models. These deviations represent a measure of the role that elastic energy plays in the rotational deformation of Earth.

The numerical values of the coefficients $f_0(a_1)$, $f_2(a_1)$, $f_4(a_1)$, and $f_6(a_1)$ provide the equation of the equilibrium configuration up to the sixth harmonic. This surface can be taken as a reference surface in gravimetric work in conjunction with the Stokes formula, which was elaborated upon in the previous chapter.

2.08 DETERMINATION OF THE EXTERIOR POTENTIAL BY METHODS OF CELESTIAL MECHANICS

We shall conclude this chapter by considering a problem of satellite geodesy: the determination of the coefficients $K_2(a_1)$, $K_4(a_1)$, and $K_6(a_1)$ of Earth's exterior potential by the tracking of artificial Earth satellites. An extensive literature exists on this topic; a good account can be found in Kaula (1966). We elaborate here on a procedure that makes use of the secular variations of the orbital elements obtainable from the Lagrangian perturbation equations.

The closed orbit of a satellite around a homogeneous, spherical Earth should be a Keplerian ellipse characterized by its six orbital elements. The real Earth causes the orbit of the satellite to deviate from an ellipse. At each instant this orbit can be approximated by an osculating ellipse; the time variation of the orbital elements of all these ellipses satisfies a fundamental

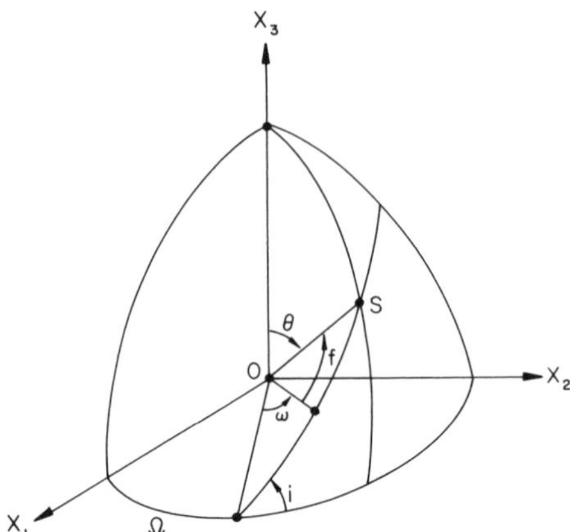

Fig. 2.3 Elements of Keplerian elliptical orbit.

system of equations, first derived by Lagrange. Consider an orthogonal reference frame $(O; X_1, X_2, X_3)$ with origin O at the centroid of the Earth, the $(X_1 X_2)$-plane coinciding with the equatorial plane, the X_1-axis pointing toward the first point of Aries, and the X_3-axis being Earth's rotational axis. The six orbital elements of a Keplerian ellipse (see Fig. 2.3) are denoted as follows:

$$\begin{aligned} \Omega &= \text{the longitude of the node,} \\ i &= \text{the inclination of the orbital plane to the} \\ &\quad \text{equatorial plane,} \\ \omega &= \text{the argument of perigee,} \\ e &= \text{the first eccentricity of the ellipse } (e < 1), \\ A &= \text{the semimajor axis,} \\ \chi &= \text{the value of the mean anomaly at time } t = 0. \end{aligned} \tag{104}$$

The mean anomaly M is expressed as

$$M = n(t - \tau) = nt + \chi, \tag{105}$$

where n is the mean angular motion and τ is the time of passage of the satellite through perigee. Other quantities that will have a bearing on the following discussion are $\mu = Gm_1$, the gravitational parameter; and the angle f, called true anomaly (see Fig. 2.3).

The Lagrangian perturbation equations of celestial mechanics express the time derivatives of the orbital elements in terms of the orbital elements themselves and of the partial derivatives of Earth's exterior potential V with respect to the same six elements. They are (see Smart, 1953)

$$
\begin{aligned}
\frac{d\Omega}{dt} &= \frac{(\mu A)^{-1/2}}{(1-e^2)^{1/2}\sin i}\frac{\partial V}{\partial i},\\
\frac{di}{dt} &= \frac{(\mu A)^{-1/2}}{(1-e^2)^{1/2}}\left[(\cot i)\frac{\partial V}{\partial \omega} - (\operatorname{cosec} i)\frac{\partial V}{\partial \Omega}\right],\\
\frac{d\omega}{dt} &= (\mu A)^{-1/2}\left[\frac{(1-e^2)^{1/2}}{e}\frac{\partial V}{\partial e} - \frac{\cot i}{(1-e^2)^{1/2}}\frac{\partial V}{\partial i}\right],\\
\frac{de}{dt} &= \frac{(\mu A)^{-1/2}}{e}\left[(1-e^2)\frac{\partial V}{\partial \chi} - (1-e^2)^{1/2}\frac{\partial V}{\partial \omega}\right],\\
\frac{dA}{dt} &= 2\left(\frac{\mu}{A}\right)^{-1/2}\frac{\partial V}{\partial \chi},\\
\frac{d\chi}{dt} &= -(\mu A)^{-1/2}\left[\frac{1-e^2}{e}\frac{\partial V}{\partial e} + 2A\frac{\partial V}{\partial A}\right].
\end{aligned}
\tag{106}
$$

The potential V which appears in these equations is the sum of even-order harmonics V_{2j}, which, from previous considerations, can be written

$$V_{2j}(r, v; a_1) = A_j\left(\frac{A}{r}\right)^{2j+1} P_{2j}(v), \qquad j = 1, 2, 3, \ldots . \tag{107}$$

Here

$$A_j = \frac{\mu}{a_1} K_{2j}(a_1)\left(\frac{a_1}{A}\right)^{2j+1} \tag{108}$$

and the P are Legendre polynomials. From Fig. 2.3 we can solve a right spherical triangle to get

$$v = \cos\theta = (\sin i)\sin(\omega + f), \tag{109}$$

which affords a transition from the colatitude θ of the satellite to its true anomaly f.

We must use the Lagrangian equations to obtain an expression for the secular variation of the elements; let it be noted here that by secular variation we mean a variation which is nonperiodic and monotonically increasing with time. For that purpose, the following steps must be taken: (1) express the potential V in terms of time through the mean anomaly M; (2) take the partial derivatives of the potential with respect to the prescribed elements; and (3) isolate those terms appearing in the right-hand side of the equation which are independent of time.

This last step is justified by the form of Eq. (106); only constant terms in the right-hand side will give rise by integration to a quantity exhibiting a secular character; all other terms have the time variable appearing through a periodic function and will yield periodic terms by integration. It will be shown that it is permissible and convenient in dealing with certain orbital elements to invert the order of the last two operations.

From the addition theorem of Legendre polynomials and Eq. (109),

$$P_{2j}(v) = \sum_{k=0}^{j} B_{jk} P_{2j}^{2k}(\cos i)\cos[2k(f+\eta)], \tag{110}$$

where

$$\eta = \omega - (\pi/2),$$

$$B_{jk} = (2-\delta_{k0})\frac{(2j-2k)!}{(2j+2k)!}P_{2j}^{2k}(0),$$

δ_{k0} being a Kronecker delta; and

$$P_n^m(x) = (1-x^2)^{m/2}\frac{d^m P_n}{dx^m}$$

are the Ferrers Legendre functions of the first kind, which have already been introduced. Note that account has already been taken in writing Eq. (110) of the fact that

$$P_n^m(0) = 0$$

when $n - m$ is an odd number.

The next question to be attacked is how to express certain factors containing powers of the radius vector multiplied by the cosine of the true anomaly in terms of the time variable, or, equivalently, of the mean anomaly. This can be achieved very concisely by means of infinite expansions first derived by Plummer (1960):

$$(r/A)^m \cos[k(f+\eta)] = \sum_{h=-\infty}^{\infty} f_h(m,k;\epsilon)\cos[(h+k)M + k\eta]. \tag{111}$$

These are valid for integral values of m and k; here $\epsilon = e/2$. The coefficients f_h have the property that

$$f_{-h}(m,k;\epsilon) = f_h(m,-k;\epsilon).$$

This enables one to consider only those terms corresponding to the positive values of the subscript h in the above summation. They can also be expressed

Table 2.10 Expansion of the Elliptical Motion up to the Sixth Power of the Eccentricity

$$f_{0,0}(m,k)=1;\; f_{0,2}(m,k)=(m+1)\,m-4k^2;$$

$$f_{0,4}(m,k)=\frac{1}{4}[(m+1)\,m(m-1)\,(m-2)-8m^2k^2+16k^4-9k^2];$$

$$f_{0,6}(m,k)=\frac{1}{36}[(m+1)\,m(m-1)\,(m-2)\,(m-3)\,(m-4)+mk(-12m^3k+48mk^3+39m^2k-48k^3-84mk+39k)-64k^6+196k^4-172k^2];$$

$$f_{1,0}(m,k)=-m+2k;\; f_{1,2}(m,k)=\frac{1}{2}[-m^3+m^2+3m+mk(2m+4k+5)-8k^3-10k^2-2k];$$

$$f_{1,4}(m,k)=\frac{1}{12}[-m^5+6m^4-5m^3-8m^2+3m+mk(2m^3+8m^2k-16mk^2-16k^3+2m^2-36mk-24k^2-12m-9k-10)$$
$$+32k^5+80k^4+26k^3-12k^2+10k];$$

$$f_{2,0}(m,k)=\frac{1}{2}(m^2-3m-4mk+4k^2+5k);$$

$$f_{2,2}(m,k)=\frac{1}{6}[m^4-6m^3-m^2+22m+mk(-4m^2+16k^2+3m+48k+47)-16k^4-60k^3-64k^2-22k];$$

$$f_{2,4}(m,k)=\frac{1}{48}[m^6-13m^5+41m^4+5m^3-58m^2-72m+mk(-4m^4-4m^3k+32m^2k^2-16mk^3-64k^4+17m^3+112m^2k-136mk^2-304k^3+66m^2$$
$$-255mk-452k^2-115m-399k-344)+64k^6+400k^5+788k^4+593k^3+314k^2+136k];$$

$$f_{3,0}(m,k)=\frac{1}{6}[-m^3+9m^2-17m+mk(6m-12k-33)+8k^3+30k^2+26k];$$

$$f_{3,2}(m,k)=\frac{1}{24}[-m^5+14m^4-41m^3-68m^2+231m+mk(6m^3-8m^2k-16mk^2+48k^3-42m^2-60mk+296k^2-76m+617k+572)$$
$$-32k^5-240k^4-614k^3-648k^2-258k];$$

$$f_{4,0}(m,k)=\frac{1}{24}[m^4-18m^3+95m^2-142m+mk(-8m^2+24mk-32k^2+102m-192k-330)+16k^4+120k^3+283k^2+206k];$$

$$f_{4,2}(m,k)=\frac{1}{120}[m^6-25m^5+185m^4-255m^3-1466m^2+3096m+mk(-8m^4+20m^3k-80mk^3+128k^4+130m^3-80m^2k-640mk^2+1360k^3-340m^2$$
$$-1765mk+5280k^2-2280m+9535k+8454)-64k^6-800k^5-3740k^4-8200k^3-8588k^2-3608k];$$

$$f_{5,0}(m,k)=\frac{1}{120}[-m^5+30m^4-305m^3+1220m^2-1569m+mk(10m^3-40m^2k+80mk^2-80k^3-230m^2+660mk-840k^2+1660m-2995k-4080)$$
$$+32k^5+400k^4+1790k^3+3360k^2+2194k];$$

$$f_{6,0}(m,k)=\frac{1}{720}[m^6-45m^5+745m^4-5595m^3+18694m^2-21576m+mk(-12m^4+60m^3k-160m^2k^2+240mk^3-192k^4+435m^3-1680m^2k+3240mk^2$$
$$-3120k^3-5530m^2+15\,345mk-18\,860k^2+29\,535m-51\,615k-60\,752)+64k^6+1200k^5+8660k^4+29\,835k^3+48\,538k^2+29\,352k].$$

as a power series of ϵ (the so-called d'Alembert series):

$$f_h(m, k; \epsilon) = \epsilon^h \sum_{n=0}^{\infty} f_{h,2n}(m, k)\epsilon^{2n},$$

this expression preserves the parity of the exponents. Table 2.10, taken from Lanzano (1971), provides the coefficients $f_{h,2n}$ as polynomials in m, k for values of the indices up to $h + 2n \leqslant 6$.

To isolate those terms in the above series (111) which are independent of time, obviously we must choose, for a fixed k, $h = -k$; thus we must consider the term

$$f_{-k}(m, k; \epsilon)\cos(k\eta) = f_k(m, -k; \epsilon)\cos(k\eta).$$

The case applicable to our potential is $m = -(2j + 1)$ and $2k$ instead of k.

Before proceeding any further, let us make the following remarks: (1) in performing the partial derivative with respect to χ, which is the only element related to M, we shall have $\partial V/\partial\chi = \partial V/\partial M$, and this will entail terms like $(h + k)\sin[(h + k)M + k\eta]$; ultimately there will be no secular contribution because we must choose $h + k = 0$; (2) as far as the partial derivatives of the potential with respect to the other orbital elements that do not imply time, it is clear that we can first perform the reduction $h + k = 0$ and next take the partial derivatives. In conclusion, we shall take the partial derivatives of the following expressions,

$$A_j \sum_{k=0}^{j} B_{jk} P_{2j}^{2k}(\cos i) f_{2k}\left[-(2j + 1), -2k; \frac{e}{2}\right] \cos\left[2k\left(\omega - \frac{\pi}{2}\right)\right],$$

with respect to A, e, ω, and i. There is no dependence on Ω; the dependence on χ has already been shown to be irrelevant; the dependence on A is through the A_j [see Eq. (108)], and the dependence on e, ω, and i is quite clear from the context. If we limit the computations up to the sixth power of the eccentricity, we can use Table 2.10. The results are displayed in Table 2.11, where the secular variation of each element has been represented by the dotted element within brackets.

Notice that

$$[\dot{A}_{2j}] = 0, \qquad [\dot{e}_2] = 0,$$

$$[di_{2j}/dt] = -\frac{e}{1 - e^2}(\cot i)[\dot{e}_{2j}].$$

The secular variation of χ has been omitted because it happens to be the least

Table 2.11

Secular Variations of the Orbital Elements e, Ω, and ω ($C = \cos^2 i$)

$$[\dot{e}_n] = \left(\frac{\mu}{A^3}\right)^{1/2}\left(\frac{a_1}{L}\right)^n K_n(a_1)e(1-e^2)(1-C)E_n$$

$$E_2 = 0$$

$$E_4 = \tfrac{15}{32}(7C-1)\sin 2\omega$$

$$E_6 = -\tfrac{105}{1024}[5(2+e^2)(33C^2-18C+1)\sin 2\omega + 3e^2(1-C)(11C-1)\sin 4\omega]$$

$$[\dot{\Omega}_n] = \left(\frac{\mu}{A^3}\right)^{1/2}\left(\frac{a_1}{L}\right)^n K_n(a_1)(\cos i)F_n$$

$$F_2 = \tfrac{3}{2}$$

$$F_4 = -\tfrac{15}{16}[(1+\tfrac{3}{2}e^2)(7C-3) - e^2(7C-4)\cos 2\omega]$$

$$\begin{aligned} F_6 = \tfrac{105}{1024}[&(8+40e^2+15e^4)(33C^2-30C+5) \\ &- 5e^2(2+e^2)(99C^2-102C+19)\cos 2\omega \\ &- \tfrac{3}{2}e^4(1-C)(33C-13)\cos 4\omega] \end{aligned}$$

$$[\dot{\omega}_n] + [\dot{\Omega}_n](\cos i) = \left(\frac{\mu}{A^3}\right)^{1/2}\left(\frac{a_1}{L}\right)^n K_n(a_1)G_n$$

$$G_2 = \tfrac{3}{4}(1-3C)$$

$$G_4 = \tfrac{15}{128}[(4+3e^2)(35C^2-30C+3) + 2(2+5e^2)(1-C)(7C-1)\cos 2\omega]$$

$$\begin{aligned} G_6 = -\tfrac{105}{2048}[&(8+20e^2+5e^4)(231C^3-315C^2+105C-5) \\ &+ 5(4+22e^2+7e^4)(1-C)(33C^2-18C+1)\cos 2\omega \\ &+ \tfrac{3}{2}e^2(4+7e^2)(1-C)^2(11C-1)\cos 4\omega] \end{aligned}$$

useful in practical cases, but it could be easily calculated from what has been said. Table 2.11 gives the secular contributions in terms of the coefficients K_2, K_4, and K_6 of the harmonics in the potential.

By tracking a satellite from a network of tracking stations, one can compute the instantaneous or osculating elements of its orbit. The variation of these elements with time will provide the rates of change that appear on the left-hand sides of the above relations; these can be symbolized, e.g., as

$$[\dot{e}_n] = F(K_{2n}).$$

These can then be solved for the unknowns K_2, K_4, and K_6. Various sophisticated algorithms devised for this purpose are available in the space literature (see, e.g., Kaula, 1966); their elaboration, however, goes beyond the scope of our presentation.

Our purpose here has been to provide a clear understanding of the theoretical background that constitutes the foundation of all the numerical processes envisaged. Table 2.10 will also find application in Chapter V.

Chapter III — Elastic Deformations of a Rotating Earth Caused by Surface Loads

3.00 INTRODUCTION

We start by considering a rotating spheroidal Earth which is in hydrostatic equilibrium under the influence of self-gravitation and its rotational acceleration. We further assume that the rotational axis is fixed with respect to the spheroid and the angular velocity is constant. We next consider the elastic yielding of the various layers comprising Earth's interior to gravity and centrifugal force. This yielding produces a variation in density distribution which is obtainable from the continuity equation and generates a deformation potential which satisfies the Poisson equation.

Elaborating on the equation of angular momentum, we reach a linearized version of the Navier–Stokes equation and look for solutions in the form of spheroidal and toroidal deformations. Seismological evidence has corroborated the existence of such permanent deformations. We develop all the variables in a series of Legendre polynomials; we ignore terms due to the Coriolis acceleration, since we are assuming no relative motion with respect to the rotating spheroid; and we take into account only the effects caused by the centrifugal acceleration.

Retention of the centrifugal acceleration in the equations of motion introduces terms which consist of the product of Legendre polynomials and their derivatives. To represent the spheroidal components of such products, we have to elaborate on the well-known Adams–Neumann product formula of the Legendre polynomials. We also make use of the Clairaut equation, discussed in Chapter II, to represent the hydrodynamic contribution within the equations of motion.

Toroidal deformations do not cause any dilatation; their solution depends on a second-order differential equation which constitutes the ϕ-component of the Navier–Stokes equation.

Spheroidal deformations, on the other hand, give rise to a system of three linear second-order equations stemming from the radial and θ-component of the Navier–Stokes equation and from the Poisson equation. We reach, however, a system simultaneously relating three orders of harmonics, specifically, the harmonics of orders $n - 2$, n, and $n + 2$. The normal form of the spheroidal deformation system is a linear system of dimension 6×6.

We formulate the numerical problem as a boundary value problem: (1) regularity conditions must be valid at the center of the configuration for all the variables, and (2) three conditions must be satisfied at the outer surface; two of them depend on the loading, and the third one expresses the harmonicity of the deformation potential.

For the purpose of initiating numerical integration, we have obtained power series solutions for the equations of motion in the neighborhood of the origin. Some coefficients of the lowest powers appearing in these series are free parameters and must be used in attempting to satisfy the three boundary conditions at the outer surface. Another parameter that can play a useful role in this connection is the initial distance from the origin where the function evaluations are performed for the first time.

In order to establish the equations valid within the liquid outer core, we evaluate the limit of the equations of motion when the rigidity approaches zero. We find that the transversal component of the stress must vanish, whereas the transversal component of the displacement can be given some degree of arbitrariness. These two variables will suffer discontinuities at the interface between liquid and solid layers. Our analysis reveals that within the liquid core we are left with a linear first-order system of only four differential equations. The adoption of a rotating Earth model avoids the controversial issue of the Adams–Williamson condition. We conclude that the Brunt–Vaisala frequency for the circulation motion of a fluid is a more appropriate parameter for studying problems associated with the stratification of the liquid core.

3.01 LINEARIZED NAVIER–STOKES EQUATIONS

We assume Earth to be initially in hydrostatic equilibrium under the influence of self-gravitation and its rotational motion. This initial state of deformation shall be considered the reference state.

With respect to an inertial frame $(O; x_1, x_2, x_3)$ with origin O coinciding with Earth's centroid, the hydrostatic stress field T_{ij} and the hydrostatic pressure p_0 are related according to

$$T_{ij} = -p_0\delta_{ij}, \qquad \partial p_0/\partial x_i = \rho_0 g_{0,i}. \tag{1}$$

Here the δ are the Kronecker deltas and ρ_0 is the mass density of the reference state. The $g_{0,i}$ are the components of gravity for the same state and are obtainable from the gravity potential:

$$g_{0,i} = \partial V_0/\partial x_i. \tag{2}$$

This potential V_0 is the sum of two terms: the gravitational potential V_0^* and the rotational potential

$$V_c = \tfrac{1}{2}\omega^2 r^2 \sin^2\theta, \tag{3}$$

where ω is Earth's rotational velocity, r is the radial distance measured from the center of gravity, and θ is the colatitude. V_0 satisfies the Poisson equation

$$\nabla^2 V_0 = -4\pi G\rho_0 + 2\omega^2, \tag{4}$$

where G is the gravitational constant.

The gravitational attraction is given by

$$\gamma_0(r) = \frac{4\pi G}{r^2}\int_0^r \rho_0(r)r^2\,dr = \frac{4\pi G}{3}\,r\bar{\rho}_0(r), \tag{5}$$

where $\bar{\rho}_0$ is the mean density, defined as follows:

$$\bar{\rho}_0(r) = \frac{3}{r^3}\int_0^r \rho_0(r)r^2\,dr. \tag{6}$$

The derivative of the gravitational attraction with respect to the radial distance r, which we shall denote by a prime, can be easily shown to be

$$\gamma_0'(r) = -(2/r)\gamma_0 + 4\pi G\rho_0. \tag{7}$$

To go one step further and consider a realistic Earth model, we must take into account the effect of the elastic forces. For this purpose, we assume that the additional stress field τ_{ij} and displacement field $\mathbf{u}(u_1, u_2, u_3)$, measured from the reference state, are related as follows:

$$\tau_{ij} = \lambda\frac{\partial u_k}{\partial x_k}\delta_{ij} + \mu\left(\frac{\partial u_i}{\partial x_j} + \frac{\partial u_j}{\partial x_i}\right). \tag{8}$$

The above formula, in which use has been made of the summation convention for repeated indices, is valid for a perfectly elastic and isotropic medium. Here λ and μ are the two Lamé parameters; μ is also known as the rigidity of the material.

Since we are assuming infinitesimal elastic deformations, Eq. (8) applies to the coordinates that the points of the medium had before the occurrence of the deformation. The same assumption, however, cannot be made for the initial hydrostatic stress field. As a consequence, the total stress at the undeformed points of the medium must be written as

$$\sigma_{ij} = T_{ij} - \frac{\partial T_{ij}}{\partial x_k} u_k + \tau_{ij}. \tag{9}$$

By differentiating Eq. (1), we can represent the gradient of the hydrostatic stress, which appears in the above equation, as

$$\frac{\partial T_{ij}}{\partial x_k} = -\frac{\partial p_0}{\partial x_k} \delta_{ij} = -\rho_0 g_{0,k} \delta_{ij}. \tag{10}$$

With respect to the inertial frame already introduced, the equations representing the displacement field **u** from the reference state can be written as

$$\rho \frac{d^2 u_i}{dt^2} = \rho \frac{\partial V}{\partial x_i} + \frac{\partial \sigma_{ij}}{\partial x_j}, \tag{11}$$

where, as before, the summation convention applies to repeated indices. Here ρ is the mass density and is the sum of ρ_0, the mass density of the reference state, and ρ_1, the change in mass density due to the displacement field. ρ_1 and the components u_i of the displacement field are related by the continuity equation:

$$\rho_1 = -\frac{\partial}{\partial x_i} (\rho_0 u_i). \tag{12}$$

V is the total potential and is the sum of V_0, the potential of the reference state, and V_1, the potential of the elastically perturbed state. V_1 consists of (1) the potential V_1^* generated by the mass redistribution due to the deformation and (2) V_e, the external potential, which represents any gravitational field acting on Earth, generated by an external perturbing body.

V_1 satisfies the Poisson equation

$$\nabla^2 V_1 = -4\pi G \rho_1. \tag{13}$$

The right-hand side of this equation is contributed by V_1^* because V_e satisfies only the Laplace equation within Earth. The contribution due to V_e will be present when expressing the boundary conditions at the outermost surface.

The components of gravity are then

$$g_i = g_{0,i} + g_{1,i}, \tag{14}$$

where

$$g_{1,i} = \partial V_1 / \partial x_i. \tag{15}$$

Neglecting the product $\rho_1 g_{1,i}$ since it is of the second order in the displacements, and using Eqs. (9) and (10), we can write the fundamental equation of motion (11):

$$\rho \frac{d^2 u_i}{dt^2} = \rho_0 g_{1,i} + \rho_1 g_{0,i} + \frac{\partial}{\partial x_i}(\rho_0 g_{0,k} u_k) + \frac{\partial \tau_{ij}}{\partial x_j}. \tag{16}$$

The last term in Eq. (16), when expanded, yields

$$\begin{aligned} \frac{\partial \tau_{ij}}{\partial x_j} &= \left(\frac{\partial \lambda}{\partial x_j}\frac{\partial u_k}{\partial x_k} + \lambda \frac{\partial^2 u_k}{\partial x_j \partial x_k}\right)\delta_{ij} \\ &\quad + \mu\left(\frac{\partial^2 u_i}{\partial x_j \partial x_j} + \frac{\partial^2 u_j}{\partial x_i \partial x_j}\right) + \frac{\partial \mu}{\partial x_j}\left(\frac{\partial u_i}{\partial x_j} + \frac{\partial u_j}{\partial x_i}\right). \end{aligned} \tag{17}$$

The reduction of this expression to vectorial form can be accomplished by taking into account the following facts:

1. $\partial u_k / \partial x_k = \nabla \cdot \mathbf{u}$ is the dilatation Δ;
2. $\partial^2 u_k / \partial x_i \partial x_k = \partial^2 u_j / \partial x_i \partial x_j$ is the ith component of the gradient of the dilatation, $\nabla(\nabla \cdot \mathbf{u})$;
3. $\partial^2 u_i / \partial x_j \partial x_j$ is the ith component of the vector Laplacian $\nabla^2 \mathbf{u}$, which is known to be representable by the vector identity $\nabla^2 \mathbf{u} = \nabla(\nabla \cdot \mathbf{u}) - \nabla \times \nabla \times \mathbf{u}$;
4. Finally, one can prove that the expression

$$\frac{\partial \mu}{\partial x_j}\left(\frac{\partial u_i}{\partial x_j} + \frac{\partial u_j}{\partial x_i}\right),$$

where the summation convention applies, is the ith component of the vector

$$(\nabla \mu)(\nabla \cdot \mathbf{u}) + \nabla(\mathbf{u} \cdot \nabla \mu) + \nabla \times (\mathbf{u} \times \nabla \mu) - \mathbf{u} \nabla \mu.$$

Replacing the above vector quantities within Eqs. (16) and (17), the fundamental equation that governs the displacement vector $\mathbf{u}$ from the reference state can be written as

$$\rho \frac{d^2 \mathbf{u}}{dt^2} = \rho_0 \mathbf{g}_1 + \rho_1 \mathbf{g}_0 + \nabla(\rho_0 \mathbf{u} \cdot \mathbf{g}_0) + \mathbf{E}(\lambda, \mu), \tag{18}$$

where

$$\begin{aligned} \mathbf{E}(\lambda, \mu) \equiv{} & (\lambda + 2\mu)\nabla(\nabla \cdot \mathbf{u}) - \mu \nabla \times \nabla \times \mathbf{u} \\ & + (\nabla \lambda + \nabla \mu)(\nabla \cdot \mathbf{u}) + \nabla(\mathbf{u} \cdot \nabla \mu) + \nabla \times (\mathbf{u} \times \nabla \mu) - \mathbf{u} \nabla^2 \mu \end{aligned} \tag{19}$$

stands for the sum of terms due to the elastic parameters λ and μ.

The first three terms on the right-hand side of Eq. (18) represent, respectively, the action of the perturbed gravity field on the initial density profile, the action of the initial gravity field on the deformed density, and the contribution due to the work performed by the elastic displacement vector against the initial hydrostatic field. Equation (18) is a linearized version of the Navier–Stokes equation in which account is taken of the variation of the material's elastic parameters.

The acceleration of **u** with respect to the inertial frame is given by

$$\frac{d^2\mathbf{u}}{dt^2} = \frac{\partial^2\mathbf{u}}{\partial t^2} + 2\boldsymbol{\omega} \times \frac{\partial \mathbf{u}}{\partial t} + \frac{d\boldsymbol{\omega}}{dt} \times \mathbf{u} + (\boldsymbol{\omega} \cdot \mathbf{u})\boldsymbol{\omega} - \omega^2\mathbf{u}, \tag{20}$$

where the partial derivative symbol refers to the variation of a vector with respect to the rotating frame.

In this chapter, we are interested in ascertaining permanent deformations of a rotating Earth, so that the first and second time derivatives of **u** with respect to the rotating frame must vanish. Also, we are assuming a constant angular rotation and no oscillation for the position of the rotational axis, which is equivalent to saying that the time derivative of $\boldsymbol{\omega}$ vanishes. We are therefore left with the last two terms of Eq. (20), which pertain to the centrifugal acceleration.

Reverting to Eq. (18), we shall write

$$\rho_0[(\boldsymbol{\omega} \cdot \mathbf{u})\boldsymbol{\omega} - \omega^2\mathbf{u}] = \rho_0\mathbf{g}_1 + \rho_1\mathbf{g}_0 + \nabla(\rho_0\mathbf{u} \cdot \mathbf{g}_0) + \mathbf{E}(\lambda, \mu), \tag{21}$$

where the term $\rho_1\mathbf{u}$ has been neglected, being of the second order in the displacements.

This is the fundamental equation of motion, and we shall seek solutions to this equation in the form of spheroidal and toroidal deformations.

In what follows we shall assume rotational symmetry, so that the displacement vector **u** depends only on the radial distance r and the colatitude θ of the position. The spherical harmonics $Y_n(\theta, \phi)$ will then reduce to the Legendre polynomials $P_n(\cos\theta)$. We use spherical coordinates (r, θ, ϕ) with the origin at the centroid, θ being the colatitude and ϕ the longitude. Accordingly, the rotational vector is representable as

$$\boldsymbol{\omega} \equiv (\omega\cos\theta;\ -\omega\sin\theta;\ 0), \tag{22}$$

and the unperturbed gravity vector as

$$\mathbf{g}_0 \equiv \nabla V_0(r, \theta) \equiv \left(\frac{\partial V_0}{\partial r}; \frac{1}{r}\frac{\partial V_0}{\partial \theta}; 0\right). \tag{23}$$

The last of the inertia terms appearing in Eq. (21) can be developed as

$$\nabla(\rho_0\mathbf{u} \cdot \mathbf{g}_0) = (\mathbf{u} \cdot \mathbf{g}_0)\nabla\rho_0 + \rho_0\nabla(\mathbf{u} \cdot \mathbf{g}_0), \tag{24}$$

where ρ_0 is a given function of r. The above expression implies knowledge of the partial derivatives of V_0 up to and including the second order. V_0 refers to the initial hydrostatic regime; its derivatives will be expressed in the next section in terms of the rotational deformation obtainable from the Clairaut equation.

The spheroidal displacement vector **u** has components of the form

$$u = \sum_{n=0}^{\infty} U_n(r)P_n(\cos\theta), \qquad v = \sum_{n=1}^{\infty} V_n(r)\, dP_n/d\theta, \qquad w = 0; \tag{25}$$

the two sequences of unknown functions $U_n(r)$, $V_n(r)$ will be determined by imposing the conditions that the above two series satisfy the fundamental equation of motion (21).

The toroidal displacement vector **u** has components of the form

$$u = 0, \qquad v = 0, \qquad w = -\sum_{n=1}^{\infty} W_n(r)\, dP_n/d\theta; \tag{26}$$

the functions $W_n(r)$ must be determined from the equations of motion.

The left-hand side of Eq. (21) will give rise to trigonometric terms that can be expressed via the Legendre polynomials as follows:

$$\begin{gathered} 3\sin^2\theta = 2(1 - P_2), \qquad 3\cos^2\theta = 1 + 2P_2, \\ 3\sin\theta\cos\theta = -dP_2/d\theta. \end{gathered} \tag{27}$$

To complete the analytic derivation of the equations of motion, one must then express the spheroidal components of the products P_2P_n, $P_2\, dP_n/d\theta$, $P_n\, dP_2/d\theta$, $(dP_2/d\theta)(dP_n/d\theta)$ for a generic value of n. This is equivalent to writing these products as linear combinations of P_n and $dP_n/d\theta$. Also, the second derivatives of the P can be eliminated by means of the fundamental equation

$$\frac{d}{d\theta}\left(\sin\theta\,\frac{dP_n}{d\theta}\right) = -n(n+1)\sin\theta\, P_n. \tag{28}$$

3.02 ANALYTIC PRELIMINARIES

We must obtain the partial derivatives of the unperturbed potential $V_0(r, \theta)$ pertaining to the hydrostatic equilibrium with respect to both variables r and θ. For this purpose, we shall use Eq. (40) of Chapter II, which expresses this potential in terms of the mean radius a which characterizes an equipotential surface. More specifically, let

$$r = a[1 + qf_{21}(a)P_2(\cos\theta)] + \cdots \tag{29}$$

be the equation of an equipotential surface to first-order terms in the rotational parameter

$$q = \frac{\omega^2 a_1^3}{3Gm_1} = \frac{\omega^2}{4\pi G\bar{\rho}_0(a_1)}, \tag{30}$$

where a_1 is the mean radius of the outermost surface and where the nondimensional function f_{21}, a solution of the Clairaut equation, represents the spheroidal deformation. In Chapter II, the potential $V_0(r, \theta)$ was determined along an equipotential surface $r(a, \theta)$ as a function of a:

$$V_0[r(a, \theta); \theta] \equiv \psi_0(a). \tag{31}$$

We use this function $\psi_0(a)$ to calculate the required derivatives. This can be achieved most expeditiously by recalling a well-known expansion theorem due to Lagrange (see, e.g., Smart, 1953, p. 27) which states: if the variables a and r are implicitly related according to the relation

$$a = r + q\phi(a),$$

then

$$F(a) = F(r) + q\phi(r)\frac{\partial F(r)}{\partial r} + \cdots, \tag{32}$$

where F stands for a generic function.

Equation (29) reveals that

$$\phi(a) \equiv -af_{21}(a)P_2(\cos\theta),$$

and this leads to

$$\begin{aligned}\psi_0(a) &= \psi_0(r) + q[-rf_{21}(r)P_2(\cos\theta)]\frac{\partial\psi_0}{\partial r} + \cdots \\ &= 4\pi G\left[\tfrac{1}{3}r^2\bar{\rho}_0(r) + \int_r^{r_1}\rho_0(r)r\,dr\right] \\ &\quad + \tfrac{1}{3}\omega^2 r^2 A^*(r)P_2(\cos\theta) + \tfrac{1}{3}\omega^2 r^2 + \cdots.\end{aligned} \tag{33}$$

Here we have set

$$A^*(r) = \frac{\bar{\rho}_0(r)}{\bar{\rho}_0(r_1)}f_{21}(r), \tag{34}$$

a nondimensional function of r, and have used the formula

$$\bar{\rho}_0' = (3/r)(\rho_0 - \bar{\rho}_0), \tag{35}$$

where primes, as usual, denote derivatives with respect to r.

By differentiation of Eq. (33) with respect to r and θ, we get

$$\begin{aligned}
\frac{\partial V_0}{\partial r} &= -\tfrac{4}{3}\pi G r \bar{\rho}_0(r) + \tfrac{2}{3}\omega^2 r + \tfrac{1}{3}\omega^2 r B^*(r) P_2(\cos\theta), \\
\frac{\partial V_0}{\partial \theta} &= \tfrac{1}{3}\omega^2 r^2 A^*(r) \frac{dP_2}{d\theta}, \\
\frac{\partial^2 V_0}{\partial r^2} &= \tfrac{4}{3}\pi G(2\bar{\rho}_0 - 3\rho_0) + \tfrac{2}{3}\omega^2 - \tfrac{1}{3}\omega^2 C^*(r) P_2(\cos\theta), \\
\frac{\partial^2 V_0}{\partial r\,\partial \theta} &= \tfrac{1}{3}\omega^2 r B^*(r) \frac{dP_2}{d\theta}, \\
\frac{\partial^2 V_0}{\partial \theta^2} &= \tfrac{1}{3}\omega^2 r^2 A^*(r) \frac{d^2 P_2}{d\theta^2}.
\end{aligned} \tag{36}$$

The following additional notation has been used:

$$\begin{aligned}
B^*(r) &= (3D_0 + \eta_{21} - 1)A^*(r), \\
C^*(r) &= (6D_0 + 2\eta_{21} - 8 - 3r\rho_0'/\bar{\rho}_0)A^*(r).
\end{aligned} \tag{37}$$

These are nondimensional expressions which depend on the solution of the Clairaut equation. We have also set

$$D_0 = \rho_0/\bar{\rho}_0, \qquad \eta = rf'/f \tag{38}$$

and have eliminated the terms containing η' by means of the Radau equation

$$r\eta' = 6 + \eta(1 - \eta) - 6(1 + \eta)D_0 \tag{39}$$

[see Eq. (28) in Chapter II]. It is easy to verify that

$$B^* = 2A^* + r(A^*)', \qquad -C^* = B^* + r(B^*)'. \tag{40}$$

These relations can be used to generate B^* and C^* if one wants to avoid numerical differentiation of ρ_0. For computational purposes, it is appropriate to note that the limit of the three functions $A^*(r)$, $B^*(r)$, and $C^*(r)$ when r approaches zero is also zero.

We shall now express the spheroidal components for the product of two Legendre polynomials and their derivatives. We start from the well-known Adams–Neumann formula (see, e.g., Whittaker and Watson, 1952) which expresses the product of two Legendre polynomials as a sum of the same polynomials:

$$P_p(\cos\theta)P_q(\cos\theta) = \sum_{r=0}^{q} B(p, q; r)P_{p+q-2r}(\cos\theta). \tag{41}$$

Here $p \geqslant q$ are positive integers, and the B coefficients are given by

$$B(p,q;r) = \frac{A_{p-r}A_r A_{q-r}}{A_{p+q-r}} \frac{2p+2q-4r+1}{2p+2q-2r+1}, \tag{42}$$

with

$$A_r = \frac{1\cdot 3\cdot 5\cdots(2r-1)}{r!}, \qquad r = 1, 2, \ldots,$$
$$A_0 = 1. \tag{43}$$

Note that p and q appear symmetrically within B. We next differentiate Eq. (41) twice with respect to θ and eliminate the second derivatives by using Eq. (28). After some reordering of the terms, we get

$$2\frac{dP_p}{d\theta}\frac{dP_q}{d\theta} - (\cot\theta)\left[\frac{d(P_pP_q)}{d\theta} - \sum_{r=0}^{q} B\frac{dP_{p+q-2r}}{d\theta}\right]$$
$$= [p(p+1)+q(q+1)]P_pP_q - \sum_{r=0}^{q}(p+q-2r)(p+q-2r+1)BP_{p+q-2r}. \tag{44}$$

The bracketed term which appears on the left-hand side vanishes because it is a differential consequence of Eq. (41). We next proceed to transform the product P_pP_q on the right-hand side by using Eq. (41) again and obtain

$$2(dP_p/d\theta)(dP_q/d\theta)$$
$$= \sum_{r=0}^{q}[p(p+1)+q(q+1)-(p+q-2r)(p+q-2r+1)]BP_{p+q-2r}.$$

We can then write

$$(dP_p/d\theta)(dP_q/d\theta) = \sum_{r=0}^{q} C(p,q;r)P_{p+q-2r}, \tag{45}$$

where

$$C(p,q;r) = [r(2p+2q-2r+1)-pq]B(p,q;r). \tag{46}$$

Note that the coefficient C is symmetrical with respect to p and q.

Let us now turn to the product $P\,dP/d\theta$ and assume that it can be represented as

$$(dP_p/d\theta)P_q = \sum_{r=0}^{q} D(p,q;r)\,dP_{p+q-2r}/d\theta. \tag{47}$$

It is a matter then of determining an algebraic expression for D. For this purpose, we differentiate Eq. (47) once with respect to θ and eliminate the second derivatives by means of Eq. (28) and write

$$\sum_{r=0}^{q} (p+q-2r)(p+q-2r+1)DP_{p+q-2r}$$
$$= p(p+1)P_pP_q - (dP_p/d\theta)(dP_q/d\theta)$$
$$+ (\cot\theta)\left[P_q\, dP_p/d\theta - \sum_{r=0}^{q} D\, dP_{p+q-2r}/d\theta\right].$$

The term within brackets vanishes because of Eq. (47). We next replace both products on the right-hand side by means of Eqs. (41) and (45). We equate the coefficients of the polynomials of like order and get

$$(p+q-2r)(p+q-2r+1)D = p(p+1)B - C.$$

Recalling the expression for C given by Eq. (46), we can finally write

$$D(p,q;r) = \frac{p(p+1) + pq - r(2p+2q-2r+1)}{(p+q-2r)(p+q-2r+1)} B(p,q;r). \tag{48}$$

The above expression is not symmetric in p and q. Naturally, we can write

$$P_p(dP_q/d\theta) = \sum_{r=0}^{q} D(q,p;r)\, dP_{p+q-2r}/d\theta.$$

Also, since

$$(dP_p/d\theta)P_q + P_p(dP_q/d\theta) = d(P_pP_q)/d\theta,$$

we must have

$$D(p,q;r) + D(q,p;r) = B(p,q;r), \tag{49}$$

and this can be directly verified.

When $q = 0$, the summations appearing in Eqs. (41), (45), and (47) reduce to only one term, which is due to $r = 0$. The limiting values of the coefficients are

$$\begin{aligned} B(p,0;0) &= 1, \qquad & C(p,0;0) &= 0, \\ D(p,0;0) &= 1, \qquad & D(0,p;0) &= 0. \end{aligned} \tag{50}$$

We can use the previously obtained formulas to evaluate the products P_nP_2, $(dP_n/d\theta)(dP_2/d\theta)$, $P_2(dP_n/d\theta)$, and $P_n(dP_2/d\theta)$, which appear in the fundamental equations of motion for a generic value of n. Table 3.1 presents the results. The terms multiplied by the factor $1 - \delta_{1n}$, where δ is the Kronecker delta, will not appear when $n = 1$.

Table 3.1

Resolution of Products of Legendre Polynomials into Spheroidal Components

$$2P_nP_2 = \frac{3(n+1)(n+2)}{(2n+1)(2n+3)}P_{n+2} + \frac{2n(n+1)}{(2n-1)(2n+3)}P_n + \frac{3(n-1)n}{(2n-1)(2n+1)}P_{n-2}$$

$$\frac{1}{3}\frac{\partial P_n}{\partial\theta}\frac{\partial P_2}{\partial\theta} = -\frac{n(n+1)(n+2)}{(2n+1)(2n+3)}P_{n+2} + \frac{n(n+1)}{(2n-1)(2n+3)}P_n + \frac{n(n-1)(n+1)}{(2n-1)(2n+1)}P_{n-2}$$

$$2\frac{\partial P_n}{\partial\theta}P_2 = \frac{3n(n+1)}{(2n+1)(2n+3)}\frac{\partial P_{n+2}}{\partial\theta} + \frac{2(n^2+n-3)}{(2n-1)(2n+3)}\frac{\partial P_n}{\partial\theta} + \frac{3n(n+1)(1-\delta_{1n})}{(2n-1)(2n+1)}\frac{\partial P_{n-2}}{\partial\theta}$$

$$\frac{1}{3}P_n\frac{\partial P_2}{\partial\theta} = \frac{n+1}{(2n+1)(2n+3)}\frac{\partial P_{n+2}}{\partial\theta} + \frac{1}{(2n-1)(2n+3)}\frac{\partial P_n}{\partial\theta} - \frac{n(1-\delta_{1n})}{(2n-1)(2n+1)}\frac{\partial P_{n-2}}{\partial\theta}$$

3.03 SPHEROIDAL DEFORMATIONS

In the previous section, we have performed the essential analytic operations that are required to write down the fundamental equations of motion as they apply to the general nth-order harmonic expansion of the spheroidal deformation. What remains to be accomplished is to express explicitly some of the auxiliary quantities appearing in the formulation of the problem.

From Eq. (25), it follows that the dilatation Δ can be written as

$$\Delta = \nabla\cdot\mathbf{u} = \sum_{n=0}^{\infty} X_n(r)P_n(\cos\theta), \tag{51}$$

where

$$X_n = U_n' + \frac{2}{r}U_n - \frac{n(n+1)}{r}V_n \tag{52}$$

and primes denote, as usual, derivatives with respect to the radial distance.

The density change due to the deformation is subject to the continuity equation and can be written as

$$\rho_1 = -\nabla\cdot(\rho_0\mathbf{u}) = -\sum_{n=0}^{\infty}(U_n\rho_0' + X_n\rho_0)P_n(\cos\theta). \tag{53}$$

Next, we consider the potential V_1 due to the deformed state; this we assume to have a spherical harmonic expansion of the type

$$V_1(r,\theta) = \sum_{n=0}^{\infty} R_n(r)P_n(\cos\theta), \tag{54}$$

thus introducing a third sequence $R_n(r)$ of unknown functions. It follows then that

$$\mathbf{g}_1 = \nabla V_1 \equiv \{R'_n P_n; (1/r) R_n \, dP_n/d\theta; 0\}. \tag{55}$$

The Poisson equation (13) satisfied by the potential V_1 can then be written as

$$r^2 R''_n + 2r R'_n - n(n+1) R_n = 4\pi G r^2 (\rho'_0 U_n + \rho_0 X_n). \tag{56}$$

The fundamental vector equation (21) will give rise to only two independent scalar equations because the two transversal components of the equation give identical contributions. The two independent components of Eq. (21) and the scalar equation (56) constitute a set of three second-order differential equations, for each value of n, that must be satisfied by the three unknowns of our problem, which are

$$\begin{aligned} U_n(r) &= \text{radial displacement,} \\ V_n(r) &= \text{tangential displacement,} \\ R_n(r) &= \text{variation in gravitational potential.} \end{aligned} \tag{57}$$

For numerical computation purposes, we must reduce our differential system to its normal form, that is to say, to an equivalent first-order differential system, by introducing three new sets of variables. These can be chosen as follows:

$$\begin{aligned} E_n(r) &= \lambda X_n + 2\mu U'_n \\ &= (\lambda + 2\mu) U'_n + \frac{2\lambda}{r} U_n - \frac{n(n+1)}{r} \lambda V_n, \end{aligned} \tag{58}$$

which expresses the radial stress

$$\tau_{rr} = \sum_{n=0}^{\infty} E_n(r) P_n(\cos\theta); \tag{59}$$

$$F_n(r) = \mu\left(V'_n - \frac{1}{r} V_n + \frac{1}{r} U_n\right), \tag{60}$$

which represents the transversal stress

$$\tau_{r\theta} = \sum_{n=1}^{\infty} F_n(r) \frac{dP_n}{d\theta}; \tag{61}$$

and

$$H_n(r) = R'_n - 4\pi G \rho_0 U_n, \tag{62}$$

which represents the gravitational flux.

By making use of the above notation, after having performed elaborate analytic manipulations based on the results of the previous section, we are able to write down the fundamental equations of motion for a rotating Earth in the form

$$Y_n' = AY_n + \rho_0\omega^2(A_1 Y_{n-2} + A_2 Y_n + A_3 Y_{n+2}). \tag{63}$$

Here Y_n is the column vector of the nth-order harmonic unknowns, in the following order,

$$Y_n = \text{column}(U_n, V_n, R_n, E_n, F_n, H_n), \tag{64}$$

and the A are 6×6 matrices that will presently be described.

The A-matrix is given in Table 3.2 in terms of ρ_0, λ and μ and agrees with the classical results first established by Alterman *et al.* (1959).

The matrices A_i ($i = 1, 2, 3$), on the other hand, have only two nonvanishing rows (the fourth and fifth), pertaining to the components E_n and F_n of the stress. It is therefore appropriate to write down the rotational contributions, which are applicable only to the E- and F-equations, more succinctly as follows:

$$\begin{aligned} E_n' &= \text{elastic terms} + \rho_0\omega^2[C_1 U_{n-2} + C_2 U_n + C_3 U_{n+2} \\ &\quad + C_4 V_{n-2} + C_5 V_n + C_6 V_{n+2} \\ &\quad + (r/\mu)(C_7 F_{n-2} + C_8 F_n + C_9 F_{n+2})], \\ F_n' &= \text{elastic terms} + \rho_0\omega^2\Big[D_1 U_{n-2} + D_2 U_n + D_3 U_{n+2} \\ &\quad + D_4 V_{n-2} + D_5 V_n + D_6 V_{n+2} \\ &\quad + \frac{r}{\lambda + 2\mu}(D_7 E_{n-2} + D_8 E_n + D_9 E_{n+2})\Big]. \end{aligned} \tag{65}$$

The C and D are altogether eighteen nondimensional functions of the radial distance r and of the order of the harmonic in consideration; these functions depend on the density ρ_0, the mean density $\bar{\rho}_0$, the hydrostatic deformation f_{21}, and their radial derivatives; they can be represented in terms of the three quantities $A^*(r)$, $B^*(r)$, and $C^*(r)$ defined by Eqs. (34) and (37). The D alone depend also on the Lamé parameters.

Tables 3.3 and 3.4 give the explicit representations of the C-functions and D-functions, respectively.

Of the six differential equations comprising our normalized system, only two, the equations expressing the derivatives of the stress, contain rotational terms. Because of these rotational terms, each equation representing the nth-order harmonic of the stress contains also the harmonics of order $n + 2$

Table 3.2

Matrix Representation for the Linear Terms of the Navier–Stokes Equations Which Are Due to the Elastic Properties of a Nonrotating Earth

$$A = \begin{bmatrix} -\dfrac{2\lambda}{\lambda+\mu}\dfrac{1}{r} & \dfrac{\lambda n(n+1)}{\lambda+2\mu}\dfrac{1}{r} & 0 & \dfrac{1}{\lambda+2\mu} & 0 & 0 \\ -\dfrac{1}{r} & \dfrac{1}{r} & 0 & 0 & \dfrac{1}{\mu} & 0 \\ 4\pi G\rho_0 & 0 & 0 & 0 & 0 & 1 \\ \dfrac{4\mu}{r^2}\left(\dfrac{3\lambda+2\mu}{\lambda+2\mu}\right)-\dfrac{4\gamma_0\rho_0}{r} & n(n+1)\left[\dfrac{\gamma_0\rho_0}{r}-\dfrac{2\mu}{r^2}\left(\dfrac{3\lambda+2\mu}{\lambda+2\mu}\right)\right] & 0 & -\dfrac{4\mu}{\lambda+2\mu}\dfrac{1}{r} & \dfrac{n(n+1)}{r} & -\rho_0 \\ \dfrac{\gamma_0\rho_0}{r}-\dfrac{2\mu}{r^2}\left(\dfrac{3\lambda+2\mu}{\lambda+2\mu}\right) & \dfrac{2\mu}{r^2}\left[\dfrac{2n(n+1)(\lambda+\mu)}{\lambda+2\mu}-1\right] & -\dfrac{\rho_0}{r} & -\dfrac{\lambda}{\lambda+2\mu}\dfrac{1}{r} & -\dfrac{3}{r} & 0 \\ 0 & -4\pi G\rho_0 n(n+1)\dfrac{1}{r} & \dfrac{n(n+1)}{r^2} & 0 & 0 & -\dfrac{2}{r} \end{bmatrix}$$

Table 3.3

Nondimensional C-Functions Representing Contributions Due to the Rotation and the Spheroidal Shape of the Earth[a]

$$C_1(n,r) = \frac{(n-1)n}{(2n-3)(2n-1)}\left[1-(n-2)A^* + B^* + \frac{1}{2}C^*\right]$$

$$C_2(n,r) = \frac{1}{3}\frac{n(n+1)}{(2n-1)(2n+3)}(2+3A^*+2B^*+C^*)$$

$$C_3(n,r) = \frac{(n+1)(n+2)}{(2n+3)(2n+5)}\left[1+(n+3)A^* + B^* + \frac{1}{2}C^*\right]$$

$$C_4(n,r) = -\frac{(n-2)(n-1)n}{(2n-3)(2n-1)}\left[1-\frac{r\rho_0'}{\rho_0}A^* + \frac{n-3}{2}B^*\right]$$

$$C_5(n,r) = -n(n+1)\left\{\frac{2}{3}+\frac{1}{(2n-1)(2n+3)}\left[-1+\frac{r\rho_0'}{\rho_0}A^* + \frac{n^2+n+3}{3}B^*\right]\right\}$$

$$C_6(n,r) = \frac{(n+1)(n+2)(n+3)}{(2n+3)(2n+5)}\left(1-\frac{r\rho_0'}{\rho_0}A^* + \frac{n+4}{2}B^*\right)$$

$$C_7(n,r) = \frac{(n-2)(n-1)n}{(2n-3)(2n-1)}A^*$$

$$C_8(n,r) = -\frac{n(n+1)}{(2n-1)(2n+3)}A^*$$

$$C_9(n,r) = -\frac{(n+1)(n+2)(n+3)}{(2n+3)(2n+5)}A^*.$$

[a] The nondimensional parameters A^*, B^*, C^* are defined in the text.

Table 3.4

Nondimensional D-Functions Representing Contributions Due to the Rotation and the Spheroidal Shape of the Earth[a]

$$D_1(n, r) = \frac{(n-1)(1-\delta_{0n})}{(2n-3)(2n-1)}\left[1 + \left(\frac{r\rho_0'}{\rho_0} + \frac{4\mu}{\lambda + 2\mu}\right) A^* - \frac{n}{2} B^*\right]$$

$$D_2(n, r) = -\frac{2}{3} + \frac{1}{(2n-1)(2n-3)}\left[1 + \left(\frac{r\rho_0'}{\rho_0} + \frac{4\mu}{\lambda + 2\mu}\right) A^* - \frac{n(n+1)}{3} B^*\right]$$

$$D_3(n, r) = -\frac{n+2}{(2n+3)(2n+5)}\left[1 + \left(\frac{r\rho_0'}{\rho_0} + \frac{4\mu}{\lambda + 2\mu}\right) A^* + \frac{n+1}{2} B^*\right]$$

$$D_4(n, r) = -\frac{(n-2)(n-1)(1-\delta_{0n})}{(2n-3)(2n-1)}\left(1 - \frac{n\lambda + 2\mu}{\lambda + 2\mu} A^*\right)$$

$$D_5(n, r) = -\frac{1}{3} - \frac{1}{(2n-1)(2n+3)}\left[\frac{2}{3}(n^2 + n - 3) + n(n+1)\frac{4\mu}{\lambda + 2\mu} A^*\right]$$

$$D_6(n, r) = -\frac{(n+2)(n+3)}{(2n+3)(2n+5)}\left[1 + \frac{(n+1)\lambda - 2\mu}{\lambda + 2\mu} A^*\right]$$

$$D_7(n, r) = \frac{(n-1)(1-\delta_{0n})}{(2n-3)(2n-1)} A^*$$

$$D_8(n, r) = \frac{1}{(2n-1)(2n+3)} A^*$$

$$D_9(n, r) = -\frac{n+2}{(2n+3)(2n+5)} A^*$$

δ_{0n} = Kronecker delta

= 1 when $n = 0$

= 0 when $n = 1, 2, 3, \ldots$

[a] The nondimensional parameters A^*, B^*, C^* are defined in the text.

and $n - 2$ of some other variables. Thus we must deal with a system simultaneously relating three different orders of harmonics; the difficulty in its numerical treatment stems from the fact that we must make some assumptions regarding the harmonics of order higher than the order we are solving or have solved for.

Before proceeding to the numerical solution of Eq. (63), we must determine

1. the conditions that must be met in order that these equations have regular solutions at the center of the configuration $r = 0$, that is to say, solutions that are expressible as infinite power series about the center;
2. the limiting form of these equations for the case $n = 0$ (zero-order harmonic), since the nth-order harmonic depends on the harmonic of order $n - 2$;
3. the limiting form of these equations that is valid within the liquid outer core where $\mu = 0$; and
4. the conditions that must be valid at the outer most surface $r = a_1$ depending on the loading conditions.

3.04 REGULARITY CONDITIONS FOR THE SOLUTIONS AT THE CENTER WHEN $\mu \neq 0$ AND $n \neq 0$

We assume power-series expansions in the neighborhood of $r = 0$ for each of the six variables

$$U_n(r) = U_{n0} + U_{n1}r + U_{n2}r^2 + \cdots,$$
$$\vdots \tag{66}$$
$$H_n(r) = H_{n0} + H_{n1}r + H_{n2}r^2 + \cdots,$$

and replace them within both sides of Eq. (63). From a physical point of view, we must also require that the components of the displacement be zero at the origin:

$$U_{n0} = 0; \qquad V_{n0} = 0. \tag{67}$$

Since the right-hand sides of Eqs. (63) contain terms in $1/r^2$ and $1/r$, the regularity conditions will be obtained by imposing the vanishing of the coefficients of the two negative powers of r and by equating the constant terms appearing on both sides. All other positive powers of r do not yield any extra conditions since we are dealing with the limit when r approaches zero. Let us remark first that from Eq. (6) it follows that

$$\lim_{r \to 0} \gamma_0(r)/r = \frac{4\pi G}{3} \bar{\rho}_0(0) \tag{68}$$

is finite; also that the rotational terms will not give any contribution because they consist of the U and V terms.

Each equation can give rise to at most three conditions. We ultimately arrive at ten conditions among the coefficients of the variables. Five of these conditions can be immediately analyzed:

$$R_{n0} = 0, \qquad R_{n1} = H_{n0}, \qquad (n-1)(n+2)H_{n0} = 0,$$
$$U_{n1} = \frac{1}{\mu} F_{n0}, \qquad R_{n2} = \frac{3}{n(n+1)} H_{n1} + 4\pi G \rho_0 V_{n1}. \tag{69}$$

From the second and third relation of Eq. (69), we can see that when $n \neq 1$, we must have

$$H_{n0} = R_{n1} = 0, \qquad n \neq 1, \tag{70}$$

whereas when $n = 1$, we can only say that

$$R_{11} = H_{10}. \tag{71}$$

The next three conditions constitute a linear homogeneous system among F_{n0}, E_{n0}, and V_{n1}. The determinant made with the coefficients of the unknowns vanishes for $n = 2$. In this case we reach solutions of the form

$$V_{21} = \frac{1}{2\mu} F_{20}, \qquad E_{20} = 2F_{20}, \tag{72}$$

whereas when $n \neq 2$, we get the solution

$$V_{n1} = E_{n0} = F_{n0} = 0, \tag{73}$$

and this leads to $U_{n1} = 0$ when $n \neq 2$. The last two conditions of the set relate the six quantities U_{n2}, V_{n2}, E_{n1}, F_{n1}, R_{n1}, and H_{n0}. In the case $n \neq 1$, when $H_{n0} = R_{n1} = 0$, these conditions can be used to ascertain U_{n2} and V_{n2} in terms of E_{n1} and F_{n1} according to the following relations:

$$\begin{aligned} 2\mu(n-1)(n+2)V_{n2} &= 3E_{n1} - (n^2+n-8)F_{n1}, \\ 2\mu(3\lambda+2\mu)(n-1)(n+2)U_{n2} &= [(5n^2+5n-1)\lambda + 6(n^2+n-1)\mu]E_{n1} \\ &\quad - n(n+1)[(2n^2+2n-13)\lambda \\ &\quad + 2(n^2+n-5)\mu]F_{n1}. \end{aligned} \tag{74}$$

When $n = 1$, the same conditions can be used to ascertain E_{11} and F_{11} in terms of U_{12}, V_{12}, and $H_{10} = R_{11}$ according to

$$E_{11} = 2\mu\left(\frac{3\lambda+2\mu}{\lambda+4\mu}\right)(U_{12} - V_{12}) - \frac{\lambda+2\mu}{\lambda+4\mu}\rho_0 H_{10},$$
$$F_{11} = -\mu\left(\frac{3\lambda+2\mu}{\lambda+4\mu}\right)(U_{12} - V_{12}) - \frac{\mu}{\lambda+4\mu}\rho_0 H_{10}. \tag{75}$$

As a consequence of the previous discussion, we reach the following results.

A. For $n = 1$ the series expansions are

$$\begin{aligned} U_1 &= U_{12}r^2 + \cdots, \\ V_1 &= V_{12}r^2 + \cdots, \\ R_1 &= H_{10}r + \tfrac{3}{2}H_{11}r^2 + \cdots, \\ E_1 &= E_{11}r + \cdots, \\ F_1 &= F_{11}r + \cdots, \\ H_1 &= H_{10} + H_{11}r + \cdots, \end{aligned} \tag{76}$$

with H_{10}, H_{11}, H_{12}, and V_{12} arbitrarily chosen and with E_{11} and F_{11} given by Eqs. (75).

B. When $n = 2$, the series expansions must be of the form

$$\begin{aligned} U_2 &= \frac{1}{\mu}F_{20}r + U_{22}r^2 + \cdots, \\ V_2 &= \frac{1}{2\mu}F_{20}r + V_{22}r^2 + \cdots, \\ R_2 &= \frac{1}{2}\left(H_{21} + \frac{4\pi G\rho_0}{\mu}F_{20}\right)r^2 + \cdots, \\ E_2 &= 2F_{20} + E_{21}r + \cdots, \\ F_2 &= F_{20} + F_{21}r + \cdots, \\ H_2 &= H_{21}r + \cdots, \end{aligned} \tag{77}$$

with arbitrary values for F_{20}, E_{21}, F_{21}, and H_{21} and with U_{22} and V_{22} given by Eqs. (74).

C. Finally, for $n \geqslant 3$, we can write expansions of the form

$$\begin{aligned} U_n &= U_{n2}r^2 + \cdots, \\ V_n &= V_{n2}r^2 + \cdots, \\ R_n &= \frac{3}{n(n+1)}H_{n1}r^2 + \cdots, \\ E_n &= E_{n1}r + \cdots, \\ F_n &= F_{n1}r + \cdots, \\ H_n &= H_{n1}r + \cdots, \end{aligned} \tag{78}$$

with E_{n1}, F_{n1}, and H_{n1} arbitrarily chosen and U_{n2} and V_{n2} also given by Eqs. (74).

3.05 FUNDAMENTAL EQUATIONS OF MOTION FOR SOME LIMITING CASES

In order to perform the numerical integration of the fundamental equations of motion, we must be able to propagate the solutions from the center of Earth to its free surface: we must, therefore, ascertain the type of equations that are valid within the liquid outer core. Furthermore, to evaluate the spheroidal deformations as infinite harmonic expansions, we must also find the system of equations that must be integrated when $n = 0$.

We plan therefore to investigate here the fundamental equations of motion that are valid in the following three limiting cases: (1) $\mu \neq 0$, $n = 0$; (2) $\mu = 0$, $n \neq 0$; and (3) $\mu = 0$, $n = 0$.

When $\mu \neq 0$, $n = 0$, there is no transverse spheroidal component for the displacement and for the stress since $dP_0/d\theta = 0$. This translates into the fact that there exists no equation for the θ-component: both $V_0(r)$ and $F_0(r)$ are arbitrary functions. The system of fundamental equations in its normal form will not contain the equations pertaining to V_0' and F_0'. Furthermore, these two variables will not appear within the remaining four equations relating U_0, R_0, E_0, and H_0 because their coefficients vanish with n.

The solution of the equation

$$H_0' = -(2/r)H_0$$

is $H_0(r) = h/r^2$; regularity at $r = 0$ requires that the arbitrary constant h be taken equal to zero. Thus $H_0 \equiv 0$ and the remaining equations are

$$\begin{aligned}
U_0' &= -\frac{2\lambda}{\lambda + 2\mu}\frac{1}{r}U_0 + \frac{1}{\lambda + 2\mu}E_0,\\
R_0' &= 4\pi G\rho_0 U_0,\\
E_0' &= 4\left[-\frac{\gamma_0\rho_0}{r} + \frac{\mu}{r^2}\left(\frac{3\lambda + 2\mu}{\lambda + 2\mu}\right)\right]U_0\\
&\quad - \frac{4\mu}{\lambda + 2\mu}\frac{1}{r}E_0 + \rho_0\omega^2\left[C_3(0)U_2 + C_6(0)V_2 + \frac{r}{\mu}C_9(0)F_2\right].
\end{aligned} \tag{79}$$

Regularity at $r = 0$ imposes the following conditions:

$$\begin{gathered}
U_{00} = R_{01} = 0, \qquad E_{00} = (3\lambda + 2\mu)U_{01},\\
E_{01} = 4\mu\left(\frac{3\lambda + 2\mu}{\lambda + 6\mu}\right)U_{02}.
\end{gathered} \tag{80}$$

Thus the series expansions in the neighborhood of $r = 0$ can be written as

$$\begin{aligned}
U_0 &= \qquad\quad U_{01}r + U_{02}r^2 + \cdots,\\
R_0 &= R_{00} \qquad\qquad + R_{02}r^2 + \cdots,\\
E_0 &= E_{00} + E_{01}r + \cdots,
\end{aligned} \tag{81}$$

where U_{01}, U_{02}, R_{00}, and R_{02} can be chosen arbitrarily and E_{00}, E_{01}, are obtainable from Eqs. (80). Note that the arbitrary functions $V_0(r)$ and $F_0(r)$ do not appear in Eqs. (79) pertaining to $n = 0$; neither will they appear in the other possible case, that is, among the rotational terms of Eqs. (65), when $n = 2$. This is so because one can verify from Tables 3.3 and 3.4 that

$$C_4(2) = C_7(2) = D_4(2) = 0.$$

To investigate the equations of motion valid within the liquid outer core, we must evaluate the limit reached by Eqs. (63) and (65) when the rigidity μ approaches zero. The R and H equations do not contain the parameter μ and do not have to be modified. The limit of the U equation can be easily calculated.

The V equation, on the other hand, when written as

$$rF_n = \lim_{\mu \to 0} \mu(rV'_n + U_n - V_n), \tag{82}$$

reveals that for the values $r \neq 0$ it must be $F_n = 0$, whereas the terms within parentheses can be arbitrarily chosen. It is physically plausible, however, to assume $V_n = 0$ throughout the whole liquid core, that is to say, to assume the vanishing of the tangential displacement. Since $F_n = 0$, the F equation will degenerate into an algebraic equation, and this can be used as an auxiliary condition. In evaluating the limit of the E-equation, we must eliminate the three terms $(r/\mu)F_p$ for $p = n - 2$, n, $n + 2$ in favor of U_p according to Eq. (82). By taking into account all these limiting conditions, and considering for the moment only the case $n \neq 0$, we reach the following differential system:

$$\begin{aligned}
V_n &= 0, \qquad F_n = 0,\\
U'_n &= -\frac{2}{r}U_n + \frac{1}{\lambda}E_n,\\
R'_n &= 4\pi G\rho_0 U_n + H_n,\\
H'_n &= \frac{n(n+1)}{r^2}R_n - \frac{2}{r}H_n,\\
E'_n &= -4\frac{\gamma_0\rho_0}{r}U_n - \rho_0 H_n + \rho_0\omega^2[(C_1 + C_7)U_{n-2}\\
&\quad + (C_2 + C_8)U_n + (C_3 + C_9)U_{n+2}],\\
E_n &= \gamma_0\rho_0 U_n - \rho_0 R_n + r\rho_0\omega^2\Big[D_1 U_{n-2}\\
&\quad + D_2 U_n + D_3 U_{n+2} + \frac{r}{\lambda}(D_7 E_{n-2}\\
&\quad + D_8 E_n + D_9 E_{n+2})\Big].
\end{aligned} \tag{83}$$

Included in the above system is an algebraic equation. No series solutions are required for Eqs. (83) because they are valid within the outer core, and their initial conditions for a value of $r \neq 0$ are furnished by the final values of the variables at the inner core interface.

If we differentiate the algebraic equation appearing in Eqs. (83) with respect to r and combine it with the E' equation of the same set, we get, after performing a number of simplifications, a relation which can symbolically be represented as

$$\left(\frac{\gamma_0 \rho_0}{\lambda} + \frac{\rho_0'}{\rho_0}\right) E_n = \rho_0 \omega^2 \sum_p \left(U_p; \frac{r}{\lambda} E_p; \frac{r^2}{\lambda} E_p'\right), \tag{84}$$

$$p = n-2, n, n+2.$$

The vanishing of the expression

$$(\gamma_0 \rho_0/\lambda) + (\rho_0'/\rho_0) \tag{85}$$

has been known as the Adams–Williamson condition (1923). It was first associated with this problem by Longman (1963) and was afterward extensively studied by Pekeris and Accad (1972). We shall provide in Chapter IV a short account of the developments according to Pekeris. The inclusion of the rotational terms in the study of the elastic deformation of Earth not only provides a more realistic physical model but also avoids the use of the Adams–Williamson condition, whose use has been controversial.

It is appropriate here to mention that the Brunt–Vaisala frequency N for a stratified liquid is given by

$$N^2 = -\frac{\gamma_0 \rho_0}{\lambda}\left(\gamma_0 + \frac{\lambda \rho_0'}{\rho_0^2}\right) = -\frac{\gamma_0^2 \rho_0}{\lambda} \beta, \tag{86}$$

where β is the nondimensional stratification function introduced by Pekeris [see Eq. (59) in Chapter IV]. N is related to the stability in the circulation motion of the liquid layer (see, e.g., Pedlosky, 1979, p. 553).

Equation (84) can then be interpreted as providing a link between the rotation of Earth and the stability of the induced motion within the liquid layer.

To summarize, we can say that in the case $\mu = 0$, $n \neq 0$, we get only four equations since the V and F variables can be safely assumed to vanish in the liquid layer of Earth's interior.

Let us finally consider the particular case when both μ and n are zero. We must consider the limit of Eqs. (83) when n approaches zero. Because of the vanishing of C_1, C_2, C_7, and C_8 when $n = 0$, and since the H-equation, as we have seen earlier, will yield $H_0 = 0$, we are left with the following differential system:

$$
\begin{aligned}
&V_0 = F_0 = H_0 = 0, \\
&U'_0 = -\frac{2}{r} U_0 + \frac{1}{\lambda} E_0, \\
&R'_0 = 4\pi G\rho_0 U_0, \\
&E'_0 = -4\frac{\gamma_0\rho_0}{r} U_0 + \rho_0\omega^2[C_3(0) + C_9(0)]U_2.
\end{aligned} \tag{87}
$$

3.06 BOUNDARY CONDITIONS. LOVE AND LOAD NUMBERS. TOROIDAL DEFORMATIONS

In performing the numerical evaluation of the spheroidal deformations we must consider three types of boundary conditions.

1. At the center of the configuration, where $r = 0$, the solutions must be regular, that is to say, they must be representable by means of infinite power series in the radial distance. These series can be used as a first-order approximation to initiate the numerical integration of the differential system. For physical reasons and independently of the external potential taken into consideration and the boundary conditions existing at the outer surface ($r = a_1$), the displacement vector must vanish at the center ($r = 0$) and the potential must remain finite at that location; we must have $U(0) = V(0) = 0$ and $R(0)$ finite.

Another approach useful in initiating the numerical integration of the system is to postulate that for $r < r_0$, where r_0 is a small value of the radial distance, the three functions λ, μ, ρ_0 can be considered to remain constant; one can then use the existing analytic solutions to this particular problem, which consist of spherical Bessel functions of the radial distance (see, e.g., Wiggins, 1968). Of these two mentioned approaches, the series solution method has been noted to give rise to a more rapid convergence.

2. At any surface representing a discontinuity between two layers of the Earth model, the variables U, R, E, and H must be continuous across them, whereas the variables V and F can be discontinuous. This situation applies in particular to the liquid–solid interface; however, between two solid regions separated by a discontinuity, we can permit V to be continuous.

3. At the outer surface ($r = a_1$) we shall consider primarily two models.

One model consists of assuming a free surface; this gives rise to the conditions

$$E_n(a_1) = F_n(a_1) = 0, \qquad n = 0, 1, 2, \ldots . \tag{88}$$

The second model considers a localized normal loading of the outer surface which is generated by a total mass m_0 uniformly distributed atop a spherical cap of semiaperture θ_0. This generates a mass density

$$\delta(a_1, \theta_0) = m_0/2\pi a_1^2(1 - \cos\theta_0). \tag{89}$$

The spherical harmonic representation of this mass distribution is given, for $n = 0$, by

$$\delta_0(a_1, \theta_0) = \tfrac{1}{2}\int_{\cos\theta_0}^{1} \delta(a_1, \theta_0)\, dv = m_0/4\pi a_1^2 \tag{90}$$

and is independent of θ_0. For $n \geqslant 1$, the representation is obtainable as follows:

$$\begin{aligned}
\delta_n(a_1, \theta_0) &= \frac{2n+1}{2}\int_{\cos\theta_0}^{1} \delta(a_1, \theta_0) P_n(v)\, dv \\
&= \frac{(2n+1)m_0}{4\pi a_1^2(1 - \cos\theta_0)}\int_{\cos\theta_0}^{1} P_n(v)\, dv \\
&= \frac{m_0}{4\pi a_1^2(1 - \cos\theta_0)}[P_{n-1}(\cos\theta_0) - P_{n+1}(\cos\theta_0)].
\end{aligned} \tag{91}$$

In reaching the above expression, we have made use of the formulas

$$\begin{aligned}
(2n+1)P_n(v) &= \frac{dP_{n+1}}{dv} - \frac{dP_{n-1}}{dv}, \qquad n = 1, 2, 3, \ldots, \\
P_n(1) &= 1, \qquad n = 0, 1, 2, \ldots
\end{aligned} \tag{92}$$

(see e.g., Hobson, 1955, p. 17, 33).

Let us now take into consideration the situation in which the spherical cap shrinks to a point, i.e., $\theta_0 \to 0$; we reach the case of a concentrated normal mass load m_0. By using Eqs. (92), we can evaluate the limit

$$\lim_{v\to 1} \frac{P_{n-1}(v) - P_{n+1}(v)}{1 - v} = \lim_{v\to 1}\left(\frac{dP_{n+1}}{dv} - \frac{dP_{n-1}}{dv}\right) = (2n+1)P_n(1) = 2n+1,$$

and thus obtain

$$\lim_{\theta_0\to 0} \delta_n(a_1, \theta_0) = \frac{(2n+1)m_0}{4\pi a_1^2} = \delta_n(a_1) \tag{93}$$

as the representation of the loading in the limiting case. Comparison with Eq. (90) shows that the above limit applies also to $n = 0$.

At the surface of Earth, the radial and transversal components of the stress must satisfy the conditions

$$\begin{aligned} E_n(a_1) &= -\gamma_0(a_1)\delta_n(a_1) = -\frac{(2n+1)m_0}{4\pi a_1^2}\gamma_0(a_1), \\ F_n(a_1) &= 0, \qquad n = 0, 1, 2, \ldots . \end{aligned} \tag{94}$$

In either of these two models, the radial displacement U_n will induce a surface density which gives rise to a discontinuity for the values that the normal derivative of the potential V_1 takes in going from the interior to the exterior of the configuration. According to Gauss's theorem of potential theory, this discontinuity amounts to $-4\pi G[\rho_0(a_1)U_n(a_1) + \delta_n(a_1)]$. The deformation potential V_1^* is harmonic outside Earth, so that

$$\partial V_1^*/\partial r = -[(n+1)/r]V_1^*. \tag{95}$$

In general, we are also assuming that there exists an external harmonic potential V_e (i.e., tidal potential), which we represent as

$$\sum K_n(r/a_1)^n Y_n(\theta, \phi), \tag{96}$$

where the K are constants, and which satisfies the condition

$$\partial V_e/\partial r = (n/r)V_e. \tag{97}$$

Piecing all these considerations together, we easily obtain the boundary condition

$$R_n'(a_1) + \frac{n+1}{a_1}R_n(a_1) = 4\pi G\rho_0(a_1)U_n(a_1) + \frac{2n+1}{a_1}\left(K_n + \frac{Gm_0}{a_1}\right).$$

In terms of the normalized variables, this can be written as

$$H_n(a_1) + \frac{n+1}{a_1}R_n(a_1) = \frac{2n+1}{a_1}\left(K_n + \frac{Gm_0}{a_1}\right). \tag{98}$$

It is customary in the geophysical literature to introduce the load numbers as nondimensional quantities related to the solution of the spheroidal deformation. Because these solutions are necessarily functions of the boundary conditions, we can have a variety of load numbers according to the loading conditions.

In case of a free surface, these numbers are called Love numbers and are defined as

$$h_n(r) = \frac{1}{r} U_n(r; f),$$

$$l_n(r) = \frac{1}{r} V_n(r; f), \tag{99}$$

$$k_n(r) = \frac{1}{r\gamma_0(r)} R_n(r; f),$$

where we have used the variable f to represent the fact that the U, V, R variables are solutions to the boundary value problem related to a free surface. These numbers depend on the order n of the harmonic approximation and the radial distance r. Most commonly, the values for $r = a_1$ are used.

If we consider a normally loaded surface, as discussed above, these same combinations are called load numbers and denoted h', l', k':

$$h'_n(r) = \frac{1}{r} U_n(r; l),$$

$$l'_n(r) = \frac{1}{r} V_n(r; l), \tag{100}$$

$$k'_n(r) = \frac{1}{r\gamma_0(r)} R_n(r; l).$$

If we neglect the effects of Earth's rotation and study the simplified differential system defined by the A-matrix given in Table 3.2, it is possible to prove that there exist some relations between the Love and load numbers as defined above. The most interesting of these relations is the following:

$$k'_n = k_n - h_n. \tag{101}$$

Details of a proof can be found in Moens (1980). Equations (63) without the rotational terms constitute a homogeneous linear system. Its general solution can be expressed as a linear combination of six particular and independent solutions forming a fundamental matrix of the system. Any matrix with a nonvanishing determinant at $r = a_1$ can be so chosen. A study of the A-matrix reveals that the U, V, R variables have a singularity at the origin ($r = 0$), whereas the E, F, H variables are regular at that location. Since the displacement has to vanish at the center and the potential must remain finite at the same location, it is easy to see that three of the arbitrary parameters appearing in the linear combination must vanish. We must therefore make use of the remaining three arbitrary parameters; this means that the final solution must be a linear combination of three linearly independent solutions that satisfy the three outer surface boundary conditions, Eqs. (94) and (98).

We wish to conclude this chapter by providing some information on the other possible type of deformation, i.e., the toroidal deformation, represented by Eq. (26), for which only the set of unknown functions $W_n(r)$ must be determined from the equations of motion.

One can immediately verify that in the case of the toroidal deformations both the dilatation Δ and the perturbed density ρ_1 vanish. As a consequence, the Poisson equation becomes

$$r^2 R_n'' + 2rR_n' - n(n+1)R_n = 0,$$

whose solution is $Ar^n + B/r^{n+1}$.

Since we want the potential to remain finite at the center of the configuration and to vanish at infinity, we must reach the conclusion that $V_1 = 0$ and that its gradient $\mathbf{g}_1$ vanishes. The Poisson equation is identically satisfied and thus does not contribute any condition.

By simple analytic manipulations, we see that only the longitude dependent component (i.e., the ϕ-component) of the equations of motion gives a contribution and that this amounts to

$$\mu\left(W_n'' + \frac{2}{r} W_n'\right) + \mu'\left(W_n' - \frac{1}{r} W_n\right) + \left[\rho_0\omega^2 - \mu\frac{n(n+1)}{r^2}\right] W_n = 0. \quad (102)$$

If we introduce the auxiliary variable

$$Z_n = \mu\left(W_n' - \frac{1}{r} W_n\right), \quad (103)$$

which represents the radial factor of the transversal component $\tau_{r\phi}$ of the stress, then we obtain the normal form of Eqs. (102):

$$\begin{aligned} W_n' &= \frac{1}{r} W_n + \frac{1}{\mu} Z_n, \\ Z_n' &= \left[(n+2)(n-1)\frac{\mu}{r^2} - \rho_0\omega^2\right] W_n - \frac{3}{r} Z_n. \end{aligned} \quad (104)$$

The boundary conditions are

$$Z_n(a_1) = 0, \qquad Z_n(b_1) = 0, \quad (105)$$

where b_1 is the radius of the outer core; they represent the vanishing of the transversal stress at these two locations.

Let us also note that when $\mu = 0$, both W_n and Z_n must vanish.

The regularity conditions at the origin can be easily obtained.

The series expansions for $n \neq 1$ are

$$\begin{aligned} W_n &= \qquad\quad r^2 W_{n2} + \cdots, \\ Z_n &= r Z_{n1} + \cdots, \end{aligned} \tag{106}$$

with arbitrary W_{n2} and

$$Z_{n1} = \tfrac{1}{4}(n+2)(n-1)\mu W_{n2}. \tag{107}$$

For $n = 1$, one gets

$$\begin{aligned} W_1 &= r W_{11} + \cdots, \\ Z_1 &= \qquad\quad r^2 Z_{12} + \cdots, \end{aligned} \tag{108}$$

where both coefficients are arbitrary parameters. No further elaboration of the above approach will be undertaken here.

Chapter IV | Solution of the Linearized Navier–Stokes Equations

4.00 INTRODUCTION

We devote this chapter to a discussion of certain numerical and asymptotic solutions to the linearized Navier–Stokes equations which were formulated in the previous chapter.

For the sake of completeness, and also for a better understanding of the results, we start by developing an asymptotic theory for the load numbers; this is formulated by making use of the Boussinesq solution pertaining to the elastic deformations of a flat plate.

Knowledge of the asymptotic behavior of the load numbers allows us to sum the infinite series of spherical harmonics that define the spheroidal deformations. These deformations comprise the components of the displacement, a variation in the intensity of gravity, the deviation in the direction of the vertical, and the components of the strain tensor existing within the deformed surface. To elucidate the numerical process, we devote some time to defining a summability algorithm and use it to establish closed form expressions for the requisite infinite series. We also take advantage of these analytic considerations to evaluate the Stokes function that had been introduced in the discussion of the geodetic preliminaries in Chapter I.

We next deal with the dynamics of the liquid core and its interaction with the solid mantle. Following classical results due to Pekeris, we use an asymptotic expansion of the equations of motion to establish the existence of a boundary layer at the top of the liquid core within which the stress can jump from a zero to a finite value, as required by the numerical integration procedure.

We provide some numerical details about the Love numbers, their dependence on the frequency of the forcing function, and on the ellipticity, mean radius, and rigidity of the core. We also show the characteristics of the spheroidal deformations as functions of the angular separation from the loading line.

We terminate the chapter by mentioning some recent results due to Wahr which take into account the effect of Earth's rotation.

4.01 BOUSSINESQ THEORY FOR AN ELASTIC FLAT PLATE

The series solutions representing the elastic deformations of a spheroid in the neighborhood of an applied load converge rather slowly; they must also mesh with the deformations experienced by the tangent plane to the spheroid at the point of the applied load. In order to elucidate the behavior of the elastic deformations at points having small angular separations from the load point, a region where the elastic forces dominate the gravitational forces, we wish to discuss briefly the response of a nongravitating, elastic, homogeneous half-space to concentrated surface loads. This is known as the Boussinesq plate problem. In the absence of the gravity vectors $\mathbf{g}_0$ and $\mathbf{g}_1$, assuming elastic equilibrium and constancy of the Lamé parameters, Eqs. (18) and (19) in Chapter III reduce to

$$(\lambda + 2\mu)\nabla(\nabla \cdot \mathbf{u}) = \mu\nabla \times \nabla \times \mathbf{u}. \tag{1}$$

We introduce a cylindrical coordinate system $(O; z, r, \theta)$, and we assume that the elastic medium is the half-space $z \leqslant 0$ and that the only load is a vertical one acting downward at the origin. Because of the assumed axial symmetry of the loading, the displacement vector is independent of the azimuth θ and can be represented as

$$\mathbf{u} = u(z, r)\hat{\mathbf{e}}_z + v(z, r)\hat{\mathbf{e}}_r, \tag{2}$$

where the $\mathbf{e}$ with the circumflex accents denote unit vectors in the z and r directions of the reference frame.

Equation (1) is equivalent to the two scalar equations

$$\begin{aligned} \left(\frac{\partial^2 u}{\partial z^2} + \frac{1}{r}\frac{\partial^2(rv)}{\partial r\,\partial z}\right)(\lambda + 2\mu) &= \frac{\mu}{r}\frac{\partial}{\partial r}\left[r\left(\frac{\partial v}{\partial z} - \frac{\partial u}{\partial r}\right)\right], \\ \left[\frac{\partial^2 u}{\partial r\,\partial z} + \frac{\partial}{\partial r}\left(\frac{1}{r}\frac{\partial(rv)}{\partial r}\right)\right](\lambda + 2\mu) &= \mu\left(\frac{\partial^2 u}{\partial r\,\partial z} - \frac{\partial^2 v}{\partial z^2}\right). \end{aligned} \tag{3}$$

The nonvanishing components of the stress tensor, using Sokolnikoff's formulation [(1956, p. 183), Eqs. (48.15)], can be written as

$$\begin{aligned} \tau_{zz} &= \lambda(\epsilon_{rr} + \epsilon_{\theta\theta} + \epsilon_{zz}) + 2\mu\epsilon_{zz} = (\lambda + 2\mu)\frac{\partial u}{\partial z} + \frac{\lambda}{r}\frac{\partial}{\partial r}(rv), \\ \tau_{rz} &= 2\mu\epsilon_{rz} = \mu\left(\frac{\partial u}{\partial r} + \frac{\partial v}{\partial z}\right). \end{aligned} \tag{4}$$

Rather than solving the second-order differential system of Eqs. (3) for u and v and subsequently determining the stress components from Eqs. (4), we shall determine simultaneously the components u, v of the displacement vector and the components τ_{zz}, τ_{rz} of the stress tensor by solving a first-order differential system.

There is a certain advantage to using the Hankel transform with respect to the radial variable r; specifically, we shall use the Hankel transform of order zero for the u and τ_{zz} variables and the Hankel transform of order one for the v and τ_{rz} variables.

Let us recall at this stage that the Hankel transform of order n of a function $f(z, r)$ of two variables with respect to one of its variables is defined as

$$F(z, \xi) = \int_0^\infty f(z, r) J_n(\xi r) r\, dr, \tag{5}$$

where J_n is the Bessel function of order n; this can be inverted as

$$f(z, r) = \int_0^\infty F(z, \xi) J_n(\xi r) \xi\, d\xi. \tag{6}$$

Let us note that, in Eqs. (5) and (6), the ξ-variable must have the dimension of an inverse length; also that the dimension of the Hankel transform $F(z, \xi)$ is equal to the dimension of the generating function $f(z, r)$ multiplied by a squared length.

Denoting the Hankel transforms of our four unknowns by the corresponding capital letters and using Eqs. (3) and (4), we reach a linear differential system of the first order; using the matrix notation this can be written as

$$\begin{bmatrix} D & \dfrac{\lambda\xi}{\lambda + 2\mu} & -\dfrac{1}{\lambda + 2\mu} & 0 \\ -\xi & D & 0 & -\dfrac{1}{\mu} \\ 0 & 0 & D & \xi \\ 0 & -\dfrac{4\mu(\lambda + \mu)\xi^2}{\lambda + 2\mu} & -\dfrac{\lambda\xi}{\lambda + 2\mu} & D \end{bmatrix} \begin{bmatrix} U \\ V \\ T_{zz} \\ T_{rz} \end{bmatrix} = 0, \tag{7}$$

where D here denotes the derivative with respect to z.

In reaching this result we have (1) assumed that the u, v variables and their partial derivatives with respect to both z and r vanish at infinity with $1/r$ and (2) made use of the following well-known relations between two Bessel functions:

$$DJ_0 = -J_1, \qquad D(zJ_1) = zJ_0. \tag{8}$$

If we denote by $C(D)$ the square matrix appearing on the left-hand side of Eq. (7), we know that the solutions to the system will be proportional to $\exp(\lambda_i z)$, where the λ are the roots of the characteristic equation defined as $\det C(T) = 0$. We can verify that this equation is

$$(T^2 - \xi^2)^2 = 0, \tag{9}$$

leading to the two repeated roots $T = \pm\xi$. The presence of repeated roots introduces solutions of the type

$$z \exp(\pm\xi z)$$

next to solutions of the pure exponential type. To pursue this matter in more detail, we consider the adjoint matrix $C_a(T)$ defined as the transpose of the matrix made with the cofactors of each element of the original matrix $C(T)$.

$C(\pm\xi)$ is a matrix of rank <4 because $\det C(\pm\xi) = 0$; the adjoint matrix $C_a(\pm\xi)$ can then be shown to have proportional columns and rows. The theory of linear differential systems (see, e.g., Frazer *et al.*, 1938, pp. 165–168) dictates that four sets of linearly independent solutions are found as follows:

$$\{\text{any column of the matrix } C_a(\pm\xi)\} \exp(\pm\xi z)$$

provides two independent solutions, and

$$\{z \text{ times any column of the matrix } C_a(\pm\xi) + \text{the same column of the matrix } C_a'(\pm\xi)\} \exp(\pm\xi z)$$

gives two more independent solutions; here the prime indicates that we must take the derivative with respect to T and subsequently replace T by $\pm\xi$. We choose the first column of $C_a(T)$ as follows:

$$\begin{matrix} T^3 - \dfrac{\xi^2(3\lambda + 4\mu)}{\lambda + 2\mu} T \\[2ex] \xi T^2 + \dfrac{\lambda\xi^3}{\lambda + 2\mu} \\[2ex] -\dfrac{4\mu(\lambda + \mu)}{\lambda + 2\mu} \xi^4 \\[2ex] \dfrac{4\mu(\lambda + \mu)}{\lambda + 2\mu} \xi^3 T \end{matrix} \tag{10}$$

and operate on it according to the procedure specified above. We reach the following sets of linearly independent solutions:

$$\begin{bmatrix} 1 \\ \mp 1 \\ \pm 2\mu\xi \\ -2\mu\xi \end{bmatrix} \exp(\pm\xi z), \qquad \begin{bmatrix} \mp\xi z + \dfrac{\mu}{\lambda+\mu} \\ \xi z \pm \dfrac{\lambda+2\mu}{\lambda+\mu} \\ -2\mu\xi^2 z \\ \pm 2\mu\xi^2 z + 2\mu\xi \end{bmatrix} \exp(\pm\xi z). \tag{11}$$

Let us now apply the boundary conditions. The four variables u, v, τ_{zz}, τ_{rz} must vanish at $z = -\infty$; this implies that the Hankel transformed variables U, V, T_{zz}, T_{rz} must also vanish at $z = -\infty$, which means that for $\xi > 0$ the terms in $\exp(-\xi z)$ shall not appear in our solution. To combine the other two solutions given by Eq. (11), we use the facts that $\tau_{rz}(0, r) = 0$, i.e., the transversal component of the stress vanishes at the free surface $z = 0$, and that the unit normal load is concentrated within a small circle of radius $r = \alpha$, i.e.,

$$\tau_{zz}(0, r) = -1/(\pi\alpha^2) \qquad \text{for} \quad r < \alpha$$

and

$$\tau_{zz}(0, r) = 0 \qquad \text{for} \quad r > \alpha.$$

It follows that

$$T_{zz}(0, \xi) = -\frac{1}{\pi\alpha^2}\int_0^\infty J_0(r\xi)r\,dr = -\frac{1}{2\pi}\frac{2J_1(\xi\alpha)}{\xi\alpha},$$

since from Eq. (8)

$$\int zJ_0(z)\,dz = zJ_1(z).$$

The limit of T_{zz} as the radius α approaches zero is $= -1/(2\pi)$. From the conditions

$$T_{rz}(0, \xi) = 0, \qquad T_{zz}(0, \xi) = -1/(2\pi), \tag{12}$$

we obtain the solutions. The displacements can be written as

$$\begin{aligned} U(z, \xi) &= -\frac{e^{\xi z}}{4\pi\mu\xi}\left(-\xi z + \frac{\lambda+2\mu}{\lambda+\mu}\right); \\ V(z, \xi) &= -\frac{e^{\xi z}}{4\pi\mu\xi}\left(\xi z + \frac{\mu}{\lambda+\mu}\right). \end{aligned} \tag{13}$$

To revert to the original variables, we must use Eq. (6) and the following integral relations:

$$\begin{aligned}
\int_0^\infty e^{-az} J_0(bz)\,dz &= (a^2 + b^2)^{-1/2}, \\
\int_0^\infty e^{-az} z J_0(bz)\,dz &= a(a^2 + b^2)^{-3/2}, \\
\int_0^\infty e^{-az} J_1(bz)\,dz &= \frac{(a^2 + b^2)^{1/2} - a}{b(a^2 + b^2)^{1/2}}, \\
\int_0^\infty e^{-az} z J_1(bz)\,dz &= \frac{b}{(a^2 + b^2)^{3/2}};
\end{aligned} \tag{14}$$

these are valid for the Bessel and exponential functions, available in Watson [1966; see also Kratzer and Franz, 1960, pp. 297–298, formulas (96) and (100)]. Note in this connection that the second and fourth equations are obtainable by differentiating the first and third with respect to the parameter a. The results can be written as

$$\begin{aligned}
u(z, r) &= -\frac{1}{4\pi\mu R}\left(\frac{\lambda + 2\mu}{\lambda + \mu} + \frac{z^2}{R^2}\right), \\
v(z, r) &= -\frac{1}{4\pi(\lambda + \mu)r}\left[1 + \frac{z}{R} + \frac{(\lambda + \mu)r^2 z}{\mu R^3}\right],
\end{aligned} \tag{15}$$

where $R^2 = r^2 + z^2$.

We have solved a purely elastic problem due to the fact that for conventional values of the density and of the Lamé parameters, the elastic forces dominate the gravitational forces. In order to ascertain the perturbation of the gravitational potential in the Boussinesq problem just solved, we shall neglect the coupling between the elastic and gravitational forces and simply solve the Poisson equation

$$\nabla^2 \phi = -4\pi G \rho_0 (\nabla \cdot \mathbf{u})$$

in the material half-space ($z \leqslant 0$), where the displacement vector $\mathbf{u}$ is furnished by Eqs. (2) and (3), and the Laplace equation in empty space ($z > 0$).

We introduce Φ as the Hankel transform of order zero of the perturbed potential ϕ; the Poisson equation will then transform into

$$\left(\frac{\partial^2}{\partial z^2} - \xi^2\right)\Phi(z, \xi) = -4\pi G \rho_0 \left(\frac{\partial U}{\partial z} + \xi V\right). \tag{16}$$

Using the Boussinesq solution [Eq. (13)], we see that the right-hand side of Eq. (16) becomes

$$\frac{2G\rho_0}{\lambda + \mu} e^{\xi z}, \qquad z \leqslant 0,$$

and will be zero for the case of the Laplace equation. The characteristic exponents for Eq. (16) can be seen to be $\pm\xi$. The solution of the Laplace equation valid for empty space must vanish at $z = +\infty$ and must be of the type

$$\Phi_e(z, \xi) = A \exp(-\xi z),$$

whereas the solution of the Poisson equation valid within the elastic medium is of the form

$$\Phi_i(z, \xi) = \left(B + \frac{G\rho_0}{\lambda + \mu} \frac{z}{\xi}\right) \exp(\xi z).$$

Φ must be continuous at the interface $z = 0$, which gives $A = B$. If we rewrite the Poisson equation as

$$\nabla \cdot (\nabla\phi + 4\pi G\rho_0 \mathbf{u}) = 0,$$

we realize that the normal component of the vector within parentheses must be continuous at $z = 0$. In terms of the transformed variables we can then write

$$\left.\frac{\partial \Phi_e}{\partial z}\right|_{z=0} = \left.\frac{\partial \Phi_i}{\partial z}\right|_{z=0} + 4\pi G\rho_0 U_i(0, \xi), \tag{17}$$

since $U_e = 0$; this allows the determination of A. We finally obtain

$$\begin{aligned} \Phi_e(z, \xi) &= \frac{G\rho_0}{2\mu\xi^2} \exp(-\xi z) && \text{for} \quad z \geqslant 0, \\ \Phi_i(z, \xi) &= \frac{G\rho_0}{2\mu\xi^2} \left(1 + \frac{2\mu\xi z}{\lambda + \mu}\right) \exp(\xi z) && \text{for} \quad z \leqslant 0. \end{aligned} \tag{18}$$

For future applications, we should remark that the fundamental results expressed by Eqs. (13), (15), and (18) refer to the unit of applied force.

4.02 ASYMPTOTIC BEHAVIOR OF THE LOAD NUMBERS

The numerical evaluation of the spheroidal deformations of a layered spheroid requires knowledge of the load numbers of arbitrarily high order. It is imperative therefore to obtain an asymptotic theory for such numbers so as to determine the cutoff point to be used for the numerical summation. This can be achieved by making use of the Boussinesq theory of an elastic flat plate.

These considerations are motivated by the following two facts. First, for small angular separations from the loading line, the displacements and the perturbed potential of a layered spheroid have to agree with the Boussinesq solution pertaining to the tangent plane to the spheroid at the loading point. It is also clear that within a small neighborhood of the loading point, the spherical reference frame can be approximated by a cylindrical frame having the loading line as its z-axis and the tangent plane to the spheroid at the loading point as the $z = 0$ plane; the unit vectors $\hat{\mathbf{e}}_z$ and $\hat{\mathbf{e}}_r$ of the cylindrical frame are the limiting positions of the unit vectors $\hat{\mathbf{e}}_r$ and $\hat{\mathbf{e}}_\theta$, respectively, of the spherical frame.

The second point of concern is that for small values of θ and large values of n, the Legendre polynomials $P_n(\theta)$ tend to approach the Bessel function $J_0(n\theta)$. This second concern touches on the well-known fact that the solutions of the Laplace equation in spherical coordinates for large values of the radius vector r are represented by

$$r^{-n}P_n^m(\cos\theta)\exp(im\phi),$$

whereas the solutions to the same equation in cylindrical coordinates are given in terms of

$$J_m(kr)\exp(\pm kz + im\phi).$$

One should expect that for large values of r and n and small values of θ, the Legendre functions should approach the Bessel functions.

Our first step will be to establish the following result:

$$\lim_{n\to\infty} P_n[\cos(\theta/n)] = J_0(\theta). \tag{19}$$

This can be verified quite easily starting from the Laplace integral representation of the Legendre polynomials

$$\begin{aligned} P_n[\cos(\theta/n)] &= \frac{1}{\pi}\int_0^\pi [\cos(\theta/n) + i\sin(\theta/n)\cos\phi]^n\,d\phi \\ &= \frac{1}{\pi}\int_0^\pi \exp\{n\ln[\cos(\theta/n) + i\sin(\theta/n)\cos\phi]\}\,d\phi. \end{aligned}$$

The limit for the argument of the exponential function as $n\to\infty$ is $(i\theta\cos\phi)$, so that we have

$$\lim_{n\to\infty} P_n[\cos(\theta/n)] = \frac{1}{\pi}\int_0^\pi \exp(i\theta\cos\phi)\,d\phi = J_0(\theta)$$

according to a well-known integral representation of the Bessel function. We can rewrite Eq. (19) as

$$\lim_{n\to\infty} P_n(\cos\alpha_n) = J_0(n\alpha_n)$$

along the null sequence α_n provided that $n\alpha_n$ remains finite.

From Eq. (19) it follows that

$$\lim_{n\to\infty} \frac{dP_n(\cos(\theta/n))}{d(\theta/n)} = -nJ_1(\theta). \tag{20}$$

The spheroidal deformations of a layered spheroid along a surface $r = a$ due to the applied load m_0, when expressed in terms of the load numbers, can be written as

$$\mathbf{u}(a,\theta) = \frac{am_0}{m} \sum_{n=0}^{\infty} \left[\hat{\mathbf{e}}_r h'_n(a) P_n(\cos\theta) + \hat{\mathbf{e}}_\theta l'_n(a) \frac{dP_n(\cos\theta)}{d\theta} \right]. \tag{21}$$

Here m is the mass located within the radius $r = a$; the circumflex accents are used to denote unit vectors. Similarly, the perturbed potential can be written as

$$\phi(a,\theta) = -\frac{am_0}{m} \gamma_0(a) \sum_{n=0}^{\infty} k'_n(a) P_n(\cos\theta). \tag{22}$$

Our previous considerations suggest that the limit of the above two equations when θ approaches zero must exhibit terms that for large values of n can be expressed within the limiting cylindrical frame as

$$\begin{aligned} &\frac{am_0}{m} [\hat{\mathbf{e}}_z h'_n(a) J_0(n\theta) - \hat{\mathbf{e}}_r n l'_n(a) J_1(n\theta)], \\ -&\frac{am_0}{m} \gamma_0(a) k'_n(a) J_0(n\theta). \end{aligned} \tag{23}$$

For large values of n, these quantities have to coincide with the corresponding quantities stemming from the Boussinesq solution at the loading point $z = 0$ and for small values of the radial distance $r = a\theta$. This comparison is facilitated by the fact that both sets are referred to the same coordinate frame.

We can use Eqs. (13) and (18) to express both the displacement vector and the disturbed potential in terms of their Hankel transforms, noting, however, that in the limit we can set $\theta = r/a$ and we can let $\xi = n/a$ since ξ has the dimension of an inverse length.

The integrals

$$\mathbf{u}(0, a\theta) = \hat{\mathbf{e}}_z \int_0^\infty U\left(0, \frac{n}{a}\right) J_0(n\theta) \frac{n}{a^2}\, dn$$
$$+ \hat{\mathbf{e}}_r \int_0^\infty V\left(0, \frac{n}{a}\right) J_1(n\theta) \frac{n}{a^2}\, dn, \tag{24}$$
$$\phi(0, a\theta) = \int_0^\infty \Phi\left(0, \frac{n}{a}\right) J_0(n\theta) \frac{n}{a^2}\, dn,$$

can be construed as appropriate limits of infinite summations; because of this, we can equate, for large values of n the corresponding terms appearing in Eqs. (23) and (24). We get the following relations,

$$\frac{am_0}{m} h_n'(a) = \frac{n}{a^2} U\left(0, \frac{n}{a}\right) m_0 \gamma_0(a),$$
$$\frac{am_0}{m} n l_n'(a) = -\frac{n}{a^2} V\left(0, \frac{n}{a}\right) m_0 \gamma_0(a), \tag{25}$$
$$\frac{am_0}{m} \gamma_0(a) k_n'(a) = -\frac{n}{a^2} \Phi\left(0, \frac{n}{a}\right) m_0 \gamma_0(a),$$

because the expressions for U, V, and Φ relate to the unit of applied force; in our case, the applied force is $m_0 \gamma_0(a)$.

The final result can be written as follows:

$$\begin{bmatrix} h^* \\ l^* \\ k^* \end{bmatrix} = \frac{m\gamma_0}{4\pi a^2 \eta} \begin{bmatrix} -\sigma/\mu \\ 1 \\ -3\rho_0 \eta / 2\mu\bar{\rho} \end{bmatrix}, \tag{26}$$

where h^*, l^*, k^* are the limits of h_n', nl_n', nk_n' as n approaches infinity, $\bar{\rho}$ is the mean density, and we have used the notation $\sigma = \lambda + 2\mu$ and $\eta = \lambda + \mu$. All quantities in Eq. (26) are to be evaluated at the outermost surface.

The above result expresses the limit as n approaches infinity of the three load numbers associated with a layered spheroid of total mass m in terms of the density, mean density, gravity, and Lamé parameters evaluated at the topmost layer.

For the 1066A Earth model, these limits are

$$h^* = -11.35767, \qquad l^* = 3.43096, \qquad k^* = -4.70473. \tag{27}$$

4.03 POISSON SUM OF CERTAIN LEGENDRE SERIES. EVALUATION OF THE STOKES SERIES

A series of the sort

$$\sum_{n=0}^{\infty} a_n P_n(\cos\theta) \tag{28}$$

is called a Legendre series; associated with it we consider the power series

$$\sum_{n=0}^{\infty} a_n h^n P_n(\cos\theta), \tag{29}$$

which we suppose to be convergent for $|h| < 1$ to a function $f(h, \cos\theta)$. The Poisson sum of the Legendre series (28) is by definition

$$\lim_{h\to 1} f(h, \cos\theta),$$

supposing that limit exist; we write then

$$\sum_{n=0}^{\infty} a_n P_n(\cos\theta) = \lim_{h\to 1} f(h, \cos\theta). \tag{30}$$

If the power series (29) is convergent for $h = 1$, then we know by the Abel theorem that the series will converge to its Poisson sum $f(1, \cos\theta)$, which turns out to be a continuous function at $h = 1$; the converse is not true. By applying the Littlewood theorem to this case (see Hobson, 1955, p. 339), we can show that only if

$$\lim_{n\to\infty} (n^{1/2} a_n) \tag{31}$$

exists and is finite can we state that the series (28) will converge in the ordinary sense to its Poisson sum. By convergence in the ordinary sense, we understand that in considering the sequence $s_0 = a_0$, $s_1 = a_0 + a_1 P_1, \ldots,$ $s_n = a_0 + \cdots + a_n P_n$, we conclude that

$$\lim_{n\to\infty} s_n = s$$

exists and is finite.

It is possible then to compute the Poisson sum of a Legendre series which is not convergent in the ordinary sense. However, one can show that if a Legendre series is summable according to the Cesaro method or any other summability method, then its sum must coincide with the Poisson sum. This situation does not constitute any setback, in essence, because from a computational point of view, many infinite series are so slowly convergent that other algorithms (e.g., the Euler transformation) must be utilized to accelerate their rates of convergence. Let us mention, in passing, that the Cesaro summability method, in its first stage, consists of finding the limit of the arithmetic mean of the partial sums:

$$\lim_{n\to\infty} \frac{s_0 + s_1 + \cdots + s_n}{n+1}. \tag{32}$$

Higher stages of this method can be easily visualized (see Knopp, 1951, p. 467).

In the numerical evaluation of the spheroidal deformations of Earth by the Kummer method, we must ultimately evaluate the Poisson sum of certain Legendre series.

Let us start our considerations with the well-known result

$$\sum_{n=0}^{\infty} h^n P_n(\cos\theta) = (1 - 2h\cos\theta + h^2)^{-1/2}. \tag{33}$$

By taking the limit of the right-hand side as h approaches one, we find that

$$\sum_{n=0}^{\infty} P_n(\cos\theta) = (2 - 2\cos\theta)^{-1/2} = \tfrac{1}{2}\operatorname{cosec}\left(\frac{\theta}{2}\right). \tag{34}$$

Next, if we differentiate Eq. (33) with respect to h, we obtain

$$\sum_{n=0}^{\infty} nh^{n-1} P_n(\cos\theta) = -(h - \cos\theta)(1 - 2h\cos\theta + h^2)^{-3/2},$$

and in taking the limit as h approaches one, we find

$$\sum_{n=0}^{\infty} nP_n(\cos\theta) = -(1 - \cos\theta)(2 - 2\cos\theta)^{-3/2} = -\tfrac{1}{4}\operatorname{cosec}\left(\frac{\theta}{2}\right). \tag{35}$$

Similarly, if we differentiate Eq. (33) with respect to θ and take the usual limit, we find

$$\sum_{n=0}^{\infty} \frac{dP_n}{d\theta} = -\tfrac{1}{4}\cos\left(\frac{\theta}{2}\right)\operatorname{cosec}^2\left(\frac{\theta}{2}\right). \tag{36}$$

Next, from the recurrence formula (see Hobson, 1955, p. 32)

$$(v^2 - 1)\frac{dP_n}{dv} = n(vP_n - P_{n-1}),$$

where $v = \cos\theta$, we easily obtain

$$(\sin\theta)\sum_{n=1}^{\infty}\left(\frac{1}{n}\frac{dP_n}{d\theta}\right) = (\cos\theta)\sum_{n=1}^{\infty} P_n - \sum_{n=1}^{\infty} P_{n-1}.$$

Using Eq. (34) and performing some trigonometric manipulations, we find

$$\sum_{n=1}^{\infty}\left(\frac{1}{n}\frac{dP_n}{d\theta}\right) = -\frac{\cos\theta + \sin(\theta/2)}{\sin\theta}. \tag{37}$$

This formula is computationally simpler than the equivalent one used by Farrell (1972).

Differentiation of Eq. (37) with respect to θ gives

$$\sum_{n=1}^{\infty}\left(\frac{1}{n}\frac{d^2P_n}{d\theta^2}\right)=\frac{1-\sin^3(\theta/2)}{\sin^2\theta}. \tag{38}$$

These are the Poisson sums of the Legendre series which are required for the numerical evaluation of the displacements (vertical and horizontal), tilt, gravity effects, and strain tensor.

We wish now to take advantage of these results to establish the finite form of the Stokes series which appears in the fundamental gravity formula introduced in Chapter I. We must evaluate

$$\sum_{n=2}^{\infty}\frac{2n+1}{n-1}P_n(\cos\theta)=2\sum_{n=2}^{\infty}P_n(\cos\theta)+3\sum_{n=2}^{\infty}\left[\frac{1}{n-1}P_n(\cos\theta)\right]. \tag{39}$$

We know how to handle the first series on the right-hand side; to evaluate the second series, we shall use the recurrence formula (see Hobson, 1955, p. 32)

$$nP_n-(2n-1)(\cos\theta)P_{n-1}+(n-1)P_{n-2}=0.$$

More conveniently, this can be rearranged as

$$\begin{aligned}2(\cos\theta)\sum_{n=2}^{\infty}P_{n-1}+(\cos\theta)\sum_{n=2}^{\infty}\left(\frac{1}{n-1}P_{n-1}\right)-\sum_{n=2}^{\infty}P_n-\sum_{n=2}^{\infty}P_{n-2}\\ =\sum_{n=2}^{\infty}\left(\frac{1}{n-1}P_n\right).\end{aligned} \tag{40}$$

Everything is known on the left-hand side, except the second series, which is equivalent to

$$\sum_{n=1}^{\infty}\left(\frac{1}{n}P_n\right)$$

and whose evaluation we shall obtain by integrating Eq. (37). We get

$$\begin{aligned}\sum_{n=1}^{\infty}\left(\frac{1}{n}P_n\right)&=-\int\frac{\cos\theta}{\sin\theta}\,d\theta-\int\frac{d\theta}{2\cos(\theta/2)}+\text{const.}\\ &=-\ln\sin\theta-\int\frac{du}{\cos u}+\text{const.}\\ &=-\ln\sin\theta+\ln\tan\left(\frac{\pi-\theta}{4}\right)+\text{const.}\end{aligned} \tag{41}$$

The integration constant can be ascertained by evaluating the limit of Eq. (41) when θ approaches π. Since $P_n(-1) = (-1)^n$, the left-hand side of Eq. (41) gives rise to the infinite series

$$\sum_{n=1}^{\infty} \frac{(-1)^n}{n} = -1 + \tfrac{1}{2} - \tfrac{1}{3} + \cdots = -\ln 2 = \ln(\tfrac{1}{2}).$$

On the other hand, the terms on the right-hand side of Eq. (41) yield an indeterminate form, which can be resolved by means of the l'Hospital formula:

$$\lim_{\theta \to \pi} \frac{\tan[(\pi - \theta)/4]}{\sin \theta} = \lim_{\theta \to \pi} \frac{-\frac{1}{4}\sec^2[(\pi - \theta)/4]}{\cos \theta} = \frac{1}{4}.$$

Using these two limits, we find that

$$\text{const.} = \ln(\tfrac{1}{2}) - \ln(\tfrac{1}{4}) = \ln 2.$$

Equation (41) becomes

$$\begin{aligned}\sum_{n=1}^{\infty} \left(\frac{1}{n} P_n\right) &= \ln\left[\frac{2\tan[(\pi - \theta)/4]}{\sin \theta}\right] \\ &= -\ln\{[\sin(\theta/2)][1 + \sin(\theta/2)]\}. \end{aligned} \tag{42}$$

Substituting Eq. (42) into Eq. (40), we obtain after some manipulations

$$\sum_{n=2}^{\infty} \left(\frac{1}{n-1} P_n\right) = 2[\sin(\theta/2)][\sin(\theta/2) - 1] - (\cos \theta)\ln[\sin(\theta/2) + \sin^2(\theta/2)]. \tag{43}$$

We now have all the ingredients for evaluating the Stokes series; from Eq. (39), using Eqs. (34) and (43), we obtain

$$\begin{aligned}\sum_{n=2}^{\infty} \frac{2n+1}{n-1} P_n(\cos \theta) &= 2[\tfrac{1}{2}\operatorname{cosec}(\theta/2) - (1 + \cos \theta)] \\ &\quad + 6[\sin(\theta/2)][\sin(\theta/2) - 1] \\ &\quad - 3(\cos \theta)\ln[\sin(\theta/2) + \sin^2(\theta/2)] \\ &= \operatorname{cosec}(\theta/2) - 4\cos^2(\theta/2) + 6\sin^2(\theta/2) - 6\sin(\theta/2) \\ &\quad - 3(\cos \theta)\ln[\sin(\theta/2) + \sin^2(\theta/2)] \\ &= \operatorname{cosec}(\theta/2) - 5[\cos^2(\theta/2) - \sin^2(\theta/2)] \\ &\quad + [\cos^2(\theta/2) + \sin^2(\theta/2)] - 6\sin(\theta/2) \\ &\quad - 3(\cos \theta)\ln[\sin(\theta/2) + \sin^2(\theta/2)] \\ &= \operatorname{cosec}(\theta/2) - 5\cos \theta + 1 - 6\sin(\theta/2) \\ &\quad - 3(\cos \theta)\ln[\sin(\theta/2) + \sin^2(\theta/2)], \end{aligned} \tag{44}$$

which is the most common and easiest way to evaluate the Stokes series.

4.04 COMPUTATION OF THE SPHEROIDAL DEFORMATIONS OF THE EARTH

The application of external loads causes deformations of the solid Earth which can be listed as

1. a displacement vector having a normal component u and a tangential component v;
2. a tilt t^* which represents the deviation from the local vertical;
3. a variation g^* in the intensity of gravity, also known as the gravity effect; and
4. an induced strain tensor with three nonvanishing components ϵ_{rr}, $\epsilon_{\theta\theta}$, $\epsilon_{\phi\phi}$.

Let us examine how to express these quantities over the spheroidal Earth as functions of the angular distance θ from the location of the applied load. We shall get infinite series of Legendre polynomials and their derivatives whose coefficients of the various harmonics depend on the load numbers. To evaluate these infinite series we make use of the well-known Kummer transformation of series, first applied by Farrell (1972) to this specific problem, which makes use of the asymptotic values of the three load numbers.

We illustrate this procedure by evaluating the radial component $u(a, \theta)$ of the spheroidal displacement due to the application of the load of mass m_0 at a distance $r = a$ from the center:

$$\begin{aligned} u(a, \theta) &= \frac{am_0}{m} \sum_{n=0}^{\infty} h_n'(a) P_n(\cos\theta) \\ &\cong \frac{am_0}{m} \left[h^* \sum_{n=0}^{\infty} P_n(\cos\theta) + \sum_{n=0}^{N} (h_n' - h^*) P_n(\cos\theta) \right]. \end{aligned} \tag{45}$$

The second term within brackets is a finite sum that will be truncated after $n = N + 1$ terms, at which stage h_n' has reached a suitable fraction of its asymptotic value h^*. According to Eq. (34), the first series sums to $(h^*/2)\operatorname{cosec}(\theta/2)$. This expression is not convergent for $\theta = 0$, so that the values obtainable from the Boussinesq approximation for the tangent plane must be used along a suitable neighborhood of $\theta = 0$. The finite sum can converge slowly; to offset this, the Euler transformation of series or other appropriate numerical transforms can be used to accelerate the convergence of the summation. For large values of n and small values of θ, one can also approximate P_n and $dP_n/d\theta$ by means of the Bessel functions J_0 and J_1, respectively, as has been seen in Section 4.02.

In much the same way, recalling that l^* is the asymptotic value of nl'_n, we can express the tangential component v of the displacement as

$$v(a,\theta) = \frac{am_0}{m}\left[l^* \sum_{n=1}^{\infty} \frac{1}{n}\frac{dP_n}{d\theta} + \sum_{n=1}^{N} (nl'_n - l^*)\frac{1}{n}\frac{dP_n}{d\theta}\right]. \tag{46}$$

The first series is given by Eq. (37) as

$$-[\cos\theta + \sin(\theta/2)]/\sin\theta,$$

and the same remark as before can be made for the values in the neighborhood of $\theta = 0$.

Let us now investigate the nonvanishing components of the strain tensor, which are nondimensional quantities. Using Eq. (63) in Chapter III and Table 3.2, we have

$$\begin{aligned}\epsilon_{rr}(a,\theta) = u'(a,\theta) &= \sum_{n=0}^{\infty} U'_n(a)P_n(\cos\theta)\\ &= \sum_{n=0}^{\infty}\left[\frac{-2\lambda(a)}{a\sigma(a)}U_n(a) + \frac{n(n+1)\lambda(a)}{a\sigma(a)}V_n(a) + \frac{E_n(a)}{\sigma(a)}\right]P_n(\cos\theta),\end{aligned}$$

where $\sigma = \lambda + 2\mu$. We note that $E_n(a)$ is a delta function with its peak at $\theta = 0$ and vanishing everywhere else; consequently it will not give any contribution for $\theta \neq 0$. Reverting to the load numbers, we can write

$$\begin{aligned}\epsilon_{rr}(a,\theta) = &-\frac{2\lambda(a)}{a\sigma(a)}u(a,\theta)\\ &+\frac{\lambda(a)}{\sigma(a)}\frac{m_0}{m}\sum_{n=0}^{\infty} n(n+1)l'_n(a)P_n(\cos\theta).\end{aligned} \tag{47}$$

The numerical computation of this infinite series necessitates the summation

$$\begin{aligned}l^*\sum_{n=0}^{\infty}(n+1)P_n(\cos\theta) &= l^*\left[\sum_{n=0}^{\infty} nP_n(\cos\theta) + \sum_{n=0}^{\infty} P_n(\cos\theta)\right]\\ &= (l^*/4)\operatorname{cosec}(\theta/2)\end{aligned}$$

[see Eqs. (35) and (34)].

Let us examine the component

$$2\epsilon_{r\theta}(r,\theta) = \frac{1}{r}\frac{\partial u}{\partial\theta} + v' - \frac{v}{r}.$$

Expressing v' by means of Eq. (63) of Chapter III and Table 3.2 and noting that F_n vanishes at the surface, we reach the result

$$\epsilon_{r\theta}(a,\theta) = 0. \tag{48}$$

Next we can compute

$$\epsilon_{\theta\theta}(a,\theta) = \frac{u(a,\theta)}{a} + \frac{1}{a}\frac{\partial v(a,\theta)}{\partial\theta}$$

$$= \frac{u(a,\theta)}{a} + \frac{m_0}{m}\sum_{n=0}^{\infty} l'_n(a)\frac{d^2P_n}{d\theta^2}; \qquad (49)$$

this depends on the series

$$l^* \sum_{n=1}^{\infty} (1/n)(d^2P_n/d\theta^2),$$

whose sum is provided by Eq. (38).

Finally, we have

$$a\epsilon_{\phi\phi}(a,\theta) = u(a,\theta) + v(a,\theta)\cot\theta, \qquad (50)$$

whose evaluation depends on the displacement components u, v, which are already available.

Let us now analytically assess the variation in the intensity of gravity due to the applied loads. A gravimeter located on the deformed crust of the Earth, at a height u above the equipotential level $r = a$ of the original configuration, will measure an acceleration due to the following potential:

$$V(a+u,\theta) = V_0(a,\theta) + uV'_0(a,\theta) + V_1^*(a,\theta) + V_e(a,\theta).$$

Here $V_0(a,\theta)$ is the potential of the unperturbed configuration; $V_1^*(a,\theta)$ is the potential due to the redistribution of mass and depends on powers like $(1/a)^{n+1}$; and $V_e(a,\theta)$ is the potential generated by the applied load and can be expressed as a series of (a^n)-terms.

The measured acceleration is then given by

$$V'(a+u,\theta) = V'_0(a,\theta) + uV''_0(a,\theta) + (V_1^*)'(a,\theta) + V'_e(a,\theta);$$

this yields

$$-g(a+u,\theta) = -g(a,\theta) + \frac{2}{a}ug(a,\theta) - \frac{n+1}{a}V_1^*(a,\theta) + \frac{n}{a}V_e(a,\theta),$$

where use has been made of Eq. (7) in Chapter III to express g'. Representing the load deformations u and V_1^* in terms of the load numbers, we get

$$g(a,\theta) - g(a+u,\theta) = g^*(a,\theta)$$

$$= \frac{m_0}{m}g(a,\theta)\sum_{n=0}^{\infty}[2h'_n(a) - (n+1)k'_n(a) + n]P_n(\cos\theta) \qquad (51)$$

because the contribution due to V_e is given by

$$(V_e)_n = \frac{m_0}{m} ag(a, \theta)P_n(\cos\theta).$$

Finally, the component $t_1^*(\theta)$ of the tilt along the meridian section can be measured by a pendulum attached to the deformed crust of the Earth; it is expressed as the angle between the normal to the equipotential surface $V_0 + V_1^* = \text{const.}$ which is related to

$$\frac{m_0}{m}\sum_{n=0}^{\infty}(1 + k_n')(dP_n/d\theta),$$

and the geometrical normal to the deformed surface, which is

$$(1/a)(\partial r/\partial\theta) = \frac{m_0}{m}\sum_{n=0}^{\infty} h_n'(dP_n/d\theta).$$

Consequently, we can write

$$t_1^*(a, \theta) = \frac{m_0}{m}\sum_{n=0}^{\infty}[1 + k_n'(a) - h_n'(a)]\frac{dP_n(\cos\theta)}{d\theta}. \tag{52}$$

4.05 DYNAMICS OF THE LIQUID CORE

We want to study the dynamics of a liquid core and its interaction with a solid mantle. A deeper insight can be gained by considering Earth's elastic deformations as the limit of harmonic oscillations when their frequencies σ_n tend to zero. For this purpose, we shall consider solutions to the spheroidal equations of Chapter III exhibiting a harmonic dependence on time through the factor $\exp(i\sigma_n t)$. In order to simplify the treatment, we shall neglect the influence of the Earth's rotation ω as a first-order approximation.

Using the same notation as in Chapter III and considering the fact that the dynamics of the liquid core implies $\mu = 0$, we find that the three sets U_n, V_n, R_n of unknown functions satisfy the equations

$$\sigma_n^2\rho_0 U_n + \rho_0 R_n' + \gamma_0\rho_0 X_n - \rho_0(\gamma_0 U_n)' + (\lambda X_n)' = 0, \tag{53}$$

$$\sigma_n^2\rho_0 r V_n + \rho_0 R_n - \gamma_0\rho_0 U_n + \lambda X_n = 0, \tag{54}$$

$$R_n'' + \frac{2}{r}R_n' - \frac{n(n+1)}{r^2}R_n = 4\pi G(\rho_0' U_n + \rho_0 X_n). \tag{55}$$

Here, as in Chapter III, we have

$$X_n = U_n' + \frac{2}{r}U_n - \frac{n(n+1)}{r}V_n \tag{56}$$

and primes denote derivatives with respect to the radial distance. If we differentiate Eq. (54) with respect to r, subtract Eq. (53) from it, and consider that σ_n is a constant, we get the following equation:

$$\sigma_n^2[(\rho_0 r V_n)' - \rho_0 U_n] + \rho_0'(R_n - \gamma_0 U_n) - \gamma_0 \rho_0 X_n = 0.$$

By again taking Eq. (54) into consideration, we simplify the above expression to

$$\left(\gamma_0 + \frac{\lambda \rho_0'}{\rho_0^2}\right) X_n = \sigma_n^2 (V_n + r V_n' - U_n). \tag{57}$$

In the limit as $\sigma_n \to 0$, either we reach the adiabatic condition

$$\gamma_0 + (\lambda \rho_0' / \rho_0^2) = 0, \tag{58}$$

which in the geophysical literature has been known as the Adams–Williamson condition, or we must suppose the vanishing of the dilatation, $X_n = 0$.

Further progress can be achieved by introducing, according to Pekeris and Accad (1972), the nondimensional stratification function $\beta(r)$ defined by

$$\gamma_0 + \frac{\lambda \rho_0'}{\rho_0^2} = \gamma_0 \beta(r), \tag{59}$$

which is related to the Brunt–Vaisala frequency N^2 by

$$\beta = -\frac{\lambda}{\rho_0 \gamma_0^2} N^2;$$

see Eq. (86) in Chapter III.

We shall call uniformly unstable a core for which $\beta(r) > 0$ throughout the core, uniformly stable a core for which $\beta(r) < 0$, and neutral a core for which $\beta(r) \equiv 0$. In many realistic Earth models, the stratification of the liquid core has been found to change with depth from stability to instability. Consequently, the assumption of neutral stratification throughout the whole liquid core would be unjustified.

Assuming therefore that $\sigma_n \to 0$ and that $X_n = 0$ throughout the core, we can solve Eqs. (54)–(56) to express this particular solution, which we denote by U_n^*, V_n^*, R_n^*, by the set of equations

$$\begin{gathered} U_n^* = R_n^* / \gamma_0, \qquad V_n^* = \frac{1}{n(n+1)} [r(U_n^*)' + 2U_n^*], \\ (R_n^*)'' + \frac{2}{r}(R_n^*)' - \left[\frac{4\pi G \rho_0'}{\gamma_0} + \frac{n(n+1)}{r^2}\right] R_n^* = 0, \end{gathered} \tag{60}$$

where it is understood that we must choose that solution R_n^* that remains finite at the origin ($r = 0$). The difficulty with the above scheme is that only one arbitrary parameter is associated with R_n^*, so that upon allowing for the discontinuity suffered by V_n^* at the core–mantle interface ($r = b$), we end up by having to use only two arbitrary constants to satisfy the three boundary conditions at the outer surface ($r = a$).

One reasonable way out of this predicament is to justify the existence of a thin boundary layer at the very top of the liquid core: within such layer, X_n would rapidly switch from the zero value assumed within the core to a finite value X_n^*. We thus get the additional condition

$$R_n - \gamma_0 U_n + (\lambda X_n^*/\rho_0) = 0 \tag{61}$$

at $r = b$. Here b is the radius of the core and both λ and ρ_0 refer to the values at the uppermost layer of the core. In what follows we shall briefly discuss the feasibility of this boundary layer; we shall justify it as stemming from an asymptotic expansion of the equations of motion when the relative contribution of certain terms appearing in them is taken into account.

Let us discuss first an ideal model consisting of a uniform liquid core surrounded by a uniform solid mantle. With constant values for ρ_0 and λ within the liquid core, Eqs. (56) and (57) can be written as

$$rX_n = \alpha_n(V_n + rV_n' - U_n) = rU_n' + 2U_n - n(n+1)V_n. \tag{62}$$

Here we have denoted by α_n the nondimensional constant

$$\alpha_n = \sigma_n^2/A, \tag{63}$$

where

$$A = \tfrac{4}{3}\pi G\rho_0, \tag{64}$$

and this implies

$$\gamma_0(r) = rA. \tag{65}$$

We introduce the new variable $Q_n = R_n/A$, which has the dimension of a squared length; Eqs. (55) and (54) can then be written as

$$Q_n'' + \frac{2}{r}Q_n' - \frac{n(n+1)}{r^2}Q_n = 3X_n, \tag{66}$$

$$(\lambda/\rho_0 A)X_n = r(U_n - \alpha_n V_n) - Q_n. \tag{67}$$

Equations (62) and (66) are two linear differential equations in U, Q, and V which can be solved, say, for U and Q in terms of V by elementary algorithms:

$$U_n = \alpha_n V_n + (n - \alpha_n)(n + 1 + \alpha_n)I_n(r), \tag{68}$$

$$Q_n = Br^n + 3r\alpha_n I_n(r), \tag{69}$$

where

$$I_n(r) = \frac{1}{r^{2+\alpha_n}} \int_0^r r^{1+\alpha_n} V_n(r)\, dr. \tag{70}$$

Note that there is only one arbitrary constant B in the above equations, instead of the expected three, because both U and Q must remain finite at $r = 0$. When we substitute the solutions we have just found into Eq. (67), it becomes apparent that X turns out to be a function of Q alone. This is so because, according to Eq. (68), $U - \alpha V$ is expressed in terms of $I(r)$, and this in turn, according to Eq. (69), depends on Q. With this value for X in its right-hand side, Eq. (66) can then be written as

$$\frac{d^2 Q_n}{ds^2} + \frac{2}{s}\frac{dQ_n}{ds} - \frac{n(n+1)}{s^2} Q_n + \frac{\rho_0 A a^2}{\lambda}\left[4 + \alpha_n - \frac{n(n+1)}{\alpha_n}\right] Q_n = L_n s^n. \tag{71}$$

Here we have introduced a new nondimensional independent variable

$$s = r/a, \tag{72}$$

and we denote by L_n the new constant

$$L_n = -\frac{\rho_0 A a^2}{\lambda \alpha_n}(n - \alpha_n)(n + 1 + \alpha_n)(Ba^n), \tag{73}$$

which depends on the arbitrary parameter B.

Equation (71) admits of the particular solution $D_n s^n$ with

$$D_n = -\frac{(n - \alpha_n)(n + 1 + \alpha_n)}{(4 + \alpha_n)\alpha_n - n(n+1)}(Ba^n). \tag{74}$$

Its general solution is then

$$Q_n = D_n s^n + T_n,$$

where T_n is the general solution of the homogeneous equation associated with Eq. (71), which is obtained from Eq. (71) by deleting its right-hand side. We are interested in ascertaining the limiting form of T_n when σ_n^2, and consequently α_n also, approach zero. We introduce for this purpose the dimensionless constant

$$v_n^2 = \frac{\rho_0 A a^2}{\lambda \alpha_n} n(n+1) = \frac{a^2 A^2}{\lambda \sigma_n^2} \rho_0 n(n+1); \tag{75}$$

we see then that the homogeneous equation related to Eq. (71) can be written as

$$\frac{d^2}{ds^2}(sT_n) - (v_n^2 + C_n)(sT_n) = 0, \tag{76}$$

where

$$C_n = \frac{n(n+1)}{s^2} - (4 + \alpha_n)\frac{\rho_0 A a^2}{\lambda} \tag{77}$$

is a coefficient which becomes negligible compared with ν_n^2 when $\alpha_n \to 0$.

The asymptotic solution to Eq. (76) can then be written as

$$T_n = \frac{F_n}{s}\exp[\nu_n(s - s_0)] + \text{terms in } \alpha_n,$$

where $s_0 = b/a$, with b the outer radius of the liquid core, and F_n is an arbitrary constant.

The general solution to Eq. (71) is

$$Q_n = D_n s^n + \frac{F_n}{s}\exp[\nu_n(s - s_0)] + O(\alpha_n), \tag{78}$$

valid for small values of α_n. Both D_n and F_n have the dimension of a squared length.

The asymptotic behaviour of the other two variables U_n, V_n can be easily ascertained using Eq. (78). This equation proves that in the liquid core the oscillations due to long-period forcing functions (i.e., with small values of σ_n) vary as $\exp(1/\sigma_n)$ rather than linearly with σ_n^2, as an inspection of Eqs. (53) and (54) would make us expect it to be in other regions of Earth's interior.

Knowing Q_n, we can easily determine both U_n and V_n. Differentiation of Eq. (70), which defines $I_n(r)$, yields

$$V_n(s) = s\frac{dI_n(s)}{ds} + (2 + \alpha_n)I_n(s). \tag{79}$$

On the other hand, from Eqs. (69) and (78), we can write

$$\begin{aligned} Q_n &= D_n s^n + \frac{F_n}{s}\exp[\nu_n(s - s_0)] \\ &= Ba^n s^n + 3as\alpha_n I_n(s); \end{aligned}$$

this will allow us to express both $I_n(s)$ and $dI_n(s)/ds$ explicitly. V_n can then be obtained from Eq. (79):

$$\begin{aligned} V_n = {} & \frac{D_n - Ba^n}{3a\alpha_n}(n + 1 + \alpha_n)s^{n-1} + \frac{F_n}{3as^2}\exp[\nu_n(s - s_0)] \\ & + \frac{F_n \nu_n}{3a\alpha_n s}\exp[\nu_n(s - s_0)]. \end{aligned}$$

Of the two exponential terms appearing in the right-hand side of the above expression, we can safely drop the first one because, for small values of α_n, it will become negligible with respect to the second one, which happens to be multiplied by $v_n\alpha_n^{-1} \sim \alpha_n^{-3/2}$. Since we have already neglected terms due to positive powers of α_n, we must do the same in the above expression. We must consider the limit as $\alpha_n \to 0$ of the coefficient of s^{n-1}; after making use of Eq. (74), we get

$$V_n(s) = \left(\frac{B}{n} a^{n-1}\right) s^{n-1} + \frac{F_n v_n}{3a\alpha_n s} \exp[v_n(s - s_0)].$$

From Eq. (68), following similar reductions and simplifications, we reach the result

$$U_n(s) = (Ba^{n-1})s^{n-1} + \frac{F_n}{3a\alpha_n} \frac{n(n+1)}{s^2} \exp[v_n(s - s_0)].$$

Similarly, the normal stress, which in the liquid core is given by λX_n, can be evaluated from Eqs. (67)–(69) in terms of Q_n; we get

$$X_n = \frac{1}{3} \frac{v_n^2}{a^2} \frac{F_n}{s} \exp[v_n(s - s_0)].$$

If we introduce the two dimensionless, arbitrary parameters

$$Ba^{n-2} = E_1, \qquad \frac{F_n}{3a^2\alpha_n} = F_1 \tag{80}$$

and use the previous results, we can express the dimensionless variables as follows:

$$\frac{1}{a} V_n(s) = \frac{1}{n} E_1 s^{n-1} + \frac{1}{s} v_n F_1 \exp[v_n(s - s_0)], \tag{81}$$

$$\frac{1}{a} U_n(s) = E_1 s^{n-1} + \frac{n(n+1)}{s^2} F_1 \exp[v_n(s - s_0)], \tag{82}$$

$$\frac{R_n(s)}{a\gamma_0(a)} = \frac{R_n(s)}{a^2 A} = \frac{Q_n(s)}{a^2} = E_1\left(1 - \frac{\alpha_n}{n}\right)s^n + \frac{3\alpha_n F_1}{s} \exp[v_n(s - s_0)], \tag{83}$$

$$X_n(s) = \frac{\alpha_n v_n^2}{s} F_1 \exp[v_n(s - s_0)]. \tag{84}$$

The solutions (81)–(83) consist of a monomial term like $E_1 s^{n-1}$ which is regular everywhere within the interval $0 \leq s \leq s_0$ and of an exponential term; on the other hand, the solution (84) for the stress exhibits only the exponential term.

This exponential term can be visualized as giving rise to an infinitesimal boundary layer at $r = b$. In fact, for large values of v_n (i.e., small values of σ_n) and for s much less than s_0, $s - s_0$ will be negative and the exponential will provide an infinitesimal contribution. On the other hand, for larger values of s such that $s - s_0$ is comparable with the small value of σ_n, the exponential will provide a finite contribution; the smaller the value of σ_n, the smaller the width of the interval within which the contribution is non-vanishing.

We can summarize this situation by stating that the stress will vanish across most of the liquid core except for a layer of decreasing thickness located at the core boundary. In the limit as $\sigma_n \to 0$, we can visualize this stress distribution as a delta function having a finite spike of infinitesimal width located at $s = s_0$.

This mathematical description of the boundary layer is akin to the one used in aerodynamics.

As we approach the boundary layer, we have $s - s_0 \to 0$ and $\exp[v_n(s - s_0)]$ will tend to one; assuming also that $\sigma_n \to 0$, the boundary conditions that are valid at the core mantle interface can be written as

$$\begin{aligned} \frac{1}{a} U_n(b) &= E_1 s_0^{n-1} + \frac{n(n+1)}{s_0^2} F_1, \\ X_n(b) &= \frac{\alpha_n v_n^2}{s_0} F_1 = n(n+1) \frac{\rho_0 A a^2}{\lambda s_0} F_1, \\ \frac{R_n(b)}{a\gamma_0(a)} &= E_1 s_0^n, \end{aligned} \tag{85}$$

whereas V_n will be discontinuous at the interface.

The two arbitrary parameters E_1, F_1 and the amount of the jump that V_n will take in going from the liquid core to the solid mantle constitute a set of three parameters we must use to satisfy the three boundary conditions at $r = a$.

Numerical results obtained by solving the Navier–Stokes equations on a computer confirm this type of behavior at the interface between liquid core and solid mantle.

We apply the previous considerations to the limiting case when the liquid core extends all the way to the upper surface, $r = a$, that is to say, to the case of a homogeneous sphere.

The vanishing of the stress at the upper surface will give rise, through Eq. (84), to $F_1 = 0$. This means that there is no boundary layer for a homogeneous liquid sphere. We can determine the value of the constant E_1 from the other condition that must be satisfied at $r = a$:

$$R'_n - 4\pi G\rho_0 U_n + \frac{n+1}{a} R_n = (2n+1)\gamma_0(a);$$

see Eq. (98) in Chapter III.

We easily obtain

$$E_1 = \frac{2n+1}{2(n-1) - [(2n+1)/n]\alpha_n}. \tag{86}$$

We use Eqs. (81)–(83) to get the three Love numbers for a liquid homogeneous sphere:

$$\begin{aligned} h_n(a) &= \frac{U_n(a)}{a} = E_1, \\ l_n(a) &= \frac{V_n(a)}{a} = \frac{1}{n} E_1, \\ k_n(a) &= \frac{R_n(a)}{a\gamma_0(a)} = E_1\left(1 - \frac{\alpha_n}{n}\right), \end{aligned} \tag{87}$$

where the value of E_1 is given by Eq. (86).

Resonance will occur when

$$\alpha_n = \frac{\sigma_n^2}{A} = \frac{2n(n-1)}{2n+1}$$

or

$$\sigma_n^2 = \frac{2n(n-1)}{2n+1}\frac{\gamma_0(a)}{a}. \tag{88}$$

This agrees with Lamb's (1945) results.

Let us now consider a more general model in which the liquid core is a sphere of radius b with continuously varying parameters $\rho_0(r)$ and $\lambda(r)$ and we are still necessarily assuming $\mu = 0$.

Equations (56) and (57) can again be combined to yield

$$rX_n = \alpha_n(V_n + rV'_n - U_n) = rU'_n + 2U_n - n(n+1)V_n. \tag{89}$$

This equation is similar to one already encountered, namely, Eq. (62); however, here α_n is not a constant but a function of r given by

$$\alpha_n(r) = r\sigma_n^2/\gamma_0(r)\beta(r),$$

where $\beta(r)$ is the stratification function defined by Eq. (59). Equation (89) can be solved for U_n in terms of V_n;

$$U_n(r) = \alpha_n(r)V_n(r) + rI_n(r), \tag{90}$$

where

$$r^3 I_n(r) = \exp(-\tau_n)\int_0^r r\exp(\tau_n)V_n[n(n+1) - \alpha_n(1+\alpha_n) - \alpha_n' r]\,dr, \tag{91}$$

and

$$\tau_n(r) = \int_0^r [\alpha_n(r)/r]\,dr. \tag{92}$$

Using Eqs. (54) and (90), we can express the right-hand side of Eq. (55) in terms of V_n and R_n; as a result, Eq. (55) can be rewritten as

$$r^2 R_n'' + 2rR_n' + \left[4\pi G\frac{\rho_0^2 r^2}{\lambda} - n(n+1)\right]R_n = 4\pi G\frac{\rho_0^2\gamma_0\beta}{\lambda}(r^3 I_n). \tag{93}$$

Let us remark at this stage that in the coefficient of R_n appearing in the left-hand side of Eq. (93), i.e., the term $4\pi G r^2\rho_0^2/\lambda$, is according to actual Earth models considerably smaller than the $n(n+1)$ term; as a matter of fact, for $n \geqslant 3$, it can be one order of magnitude smaller. It will therefore be permissible in future considerations to subsume the former term in the latter one.

Denoting by Q_n the differential operator which constitutes the left-hand side of Eq. (93) and using Eqs. (91) and (92), we can write

$$\begin{aligned}\frac{d}{dr}\left(\frac{\lambda e^{\tau_n}}{\rho_0^2\gamma_0\beta}Q_n\right) &= 4\pi G r e^{\tau_n}V_n[n(n+1) - \alpha_n(1+\alpha_n) - r\alpha_n'] \\ &= e^{\tau_n}\frac{d}{dr}\left(\frac{\lambda Q_n}{\rho_0^2\gamma_0\beta}\right) + \frac{\alpha_n}{r}e^{\tau_n}\frac{\lambda Q_n}{\rho_0^2\gamma_0\beta}.\end{aligned} \tag{94}$$

Hence

$$\frac{d}{dr}\left(\frac{\lambda Q_n}{\rho_0^2\gamma_0\beta}\right) = 4\pi G r n(n+1)V_n + (\text{terms in } \alpha_n Q_n \text{ and } \alpha_n V_n), \tag{95}$$

where α_n is to be considered again a small parameter because we are trying to establish an asymptotic solution for $\sigma_n \to 0$. We next eliminate U_n among the three Eqs. (54), (89) and (90); after some simplifications, we reach the following result:

$$\left(\gamma_0 + \frac{\lambda\alpha_n}{r\rho_0}\right)r^3 I_n = r^2 R_n + \frac{r\lambda\alpha_n}{\rho_0}(rV_n)' + \frac{r\lambda\alpha_n}{\rho_0}\left(\frac{r\rho_0'}{\rho_0} - \alpha_n\right)V_n. \tag{96}$$

Combining Eqs. (96) with (93) will give

$$\frac{\lambda Q_n}{\beta\gamma_0\rho_0^2} = \frac{\lambda}{\beta\gamma_0\rho_0^2}\left\{r^2 R_n'' + 2rR_n' + \left[4\pi G\frac{\rho_0^2 r^2}{\lambda} - n(n+1)\right]R_n\right\}$$

$$= \frac{4\pi G}{\gamma_0 + \dfrac{\lambda\alpha_n}{r\rho_0}}\left\{r^2 R_n + \frac{r\lambda\alpha_n}{\rho_0}(rV_n)' + \frac{r\lambda\alpha_n}{\rho_0}\left(\frac{r\rho_0'}{\rho_0} - \alpha_n\right)V_n\right\}. \quad (97)$$

In comparing the relative contribution of the various terms appearing in the above equation, let us state that (1) $\beta(r)$ is of the order of 10^{-1} and (2) due to the slow variation of density with radial distance, ρ_0' can also be considered a small parameter. We can therefore neglect (1) the two terms in α_n^2 and $\alpha_n\rho_0'$ in comparison with the term in α_n appearing within the right-hand side; (2) the term $\lambda\alpha_n/r\rho_0$ in comparison with γ_0; (3) the term in $r^2 R_n$ on the right-hand side in comparison with the corresponding terms in the left, the more so since the factor β^{-1} appears on the left.

Taking these reductions into consideration, and making use of Eq. (95), we finally obtain

$$\left(\frac{\lambda Q_n}{\beta\gamma_0\rho_0^2}\right)' = 4\pi G n(n+1) rV_n = 4\pi G\left[\frac{r\lambda\alpha_n}{\rho_0\gamma_0}(rV_n)'\right]'. \quad (98)$$

We shall neglect, furthermore, the radial derivative of the function $r\lambda\alpha_n/\rho_0\gamma_0$ because it does not exhibit any appreciable variation with r and because it turns out to be proportional to σ_n^2; we get then the fundamental equation

$$(rV_n)'' = \frac{c_n^2(r)}{\sigma_n^2}(rV_n), \quad (99)$$

where

$$c_n^2(r) = \frac{n(n+1)\beta\rho_0\gamma_0^2}{r^2\lambda} \quad (100)$$

and, as usual, primes denote derivatives with respect to r. The asymptotic solution of Eq. (99) which has the large parameter c^2/σ^2 can be obtained by means of the transformation of variable defined by

$$(rV)'/(rV) = z/\sigma. \quad (101)$$

By differentiation, this gives

$$(rV)''/(rV) - [(rV)'/(rV)]^2 = z'/\sigma; \quad (102)$$

upon substituting Eqs. (99) and (101) in Eq. (102), we get

$$\sigma z' + z^2 = c^2. \tag{103}$$

We seek an asymptotic solution of the above equation of the form

$$z(r) = z_0(r) + \sigma z_1(r) + \sigma^2 z_2(r) + \cdots.$$

We can easily see that the various unknown functions $z_0, z_1, \ldots$ are obtainable from the recursive consistent relations

$$\begin{gathered} z_0^2 = c^2, \qquad z_0' + 2z_0 z_1 = 0, \qquad z_1' + 2z_0 z_2 + z_1^2 = 0, \ldots, \\ z_n' + 2z_0 z_{n+1} + \sum_{k=1}^{n} z_k z_{n+1-k} = 0. \end{gathered} \tag{104}$$

The leading term, $z_0 = c$, gives rise to the solution

$$rV = A \exp\left[-\frac{1}{\sigma}\int_r^b c(r)\,dr\right] + O(\sigma), \tag{105}$$

where $A = bV(b)$.

Using this asymptotic approximation for V, we can proceed to establish similar results for the other variables of concern to our problem. First, using Eqs. (91), (92) and (105),

$$\begin{aligned} r^3 I(r) &= e^{-\tau}\int_0^r e^{\tau}(rV)[n(n+1) - \alpha(1+\alpha) - r\alpha']\,dr \\ &\cong \frac{n(n+1)A}{c}\,\sigma \exp\left[-\frac{1}{\sigma}\int_r^b c(r)\,dr\right] + \cdots. \end{aligned} \tag{106}$$

Next, from Eqs. (90), (105), and (106),

$$U = \alpha V + rI \cong \frac{n(n+1)A}{cr^2}\,\sigma \exp\left[-\frac{1}{\sigma}\int_r^b c(r)\,dr\right] + \cdots. \tag{107}$$

Also, from Eqs. (89) and (105),

$$X = \frac{\alpha}{r}(rV)' - \frac{\alpha}{r}U \cong \frac{Ac\alpha}{r\sigma}\exp\left[-\frac{1}{\sigma}\int_r^b c(r)\,dr\right] + \cdots. \tag{108}$$

Finally, one can see, guided mostly by numerical results, that the boundary layer contribution to the R-variable consists essentially of terms in σ^2 and as such should not be considered in the first-order approximation.

To summarize these considerations on the dynamic behavior of a polytropic core, mention should be made of the fact that the boundary layer contribution just established should be added to the regular solution U^*, V^*, R^*, which corresponds to $X = 0$, $\sigma = 0$ and which is obtainable by solving Eq. (60). Returning to the nondimensional variable $s = r/a$ and introducing the notation

$$v_n^2(s) = n(n+1)\beta\rho_0\gamma_0^2/\lambda s^2 = a^2c_n^2(r), \tag{109}$$

$$s_0 = b/a, \qquad M_n(s) = \exp\left\{-\frac{1}{\sigma_n}\int_s^{s_0} v_n(s)\,ds\right\},$$

we shall write the general solution in terms of two arbitrary parameters E, F as follows:

$$\begin{aligned}
U_n &= EU_n^* + F\sigma_n M_n/s^2 v_n,\\
V_n &= EV_n^* + FM_n/n(n+1)s,\\
R_n &= ER_n^* + O(\sigma^2),\\
H_n &= R_n' - 4\pi G\rho_0 U_n = EH_n^* - 4\pi G\lambda X_n/\gamma_0,\\
\lambda X_n &= \rho_0\gamma_0 F\sigma_n M_n/s^2 v_n.
\end{aligned} \tag{110}$$

Note that in the case of stable stratification, when $\beta(r) < 0$, $v(s)$ is an imaginary quantity; in this case the exponential $M(s)$ must be replaced by

$$\sin\left[\frac{1}{\sigma}\int_s^{s_0}|v(s)|\,ds\right].$$

The expression for the stress λX consists only of the boundary layer factor; if $M(s)$ is of the exponential form ($\beta > 0$), then the stress will decrease with depth below the core surface; if $\beta < 0$, then the stress will oscillate as $\sin(1/\sigma)$.

The previous expressions contain terms in σ/v, that is to say, $\sim\sigma/\beta^{1/2}$; these terms cannot be used in the case of neutral stratification ($\beta = 0$). In this case Eq. (57) becomes

$$\beta\gamma_0 X_n = \sigma_n^2(V_n + rV_n' - U_n).$$

For $\sigma_n \neq 0$, we must have

$$(rV_n)' = U_n. \tag{111}$$

Numerical solution of the Navier–Stokes equation in this case shows a very slight variation of the stress with σ_n; this variation can be approximated linearly in terms of σ_n^2 with coefficients of the order of 10^{-5}. In the limit as $\sigma_n \to 0$ (β being always zero), the stress solution does not depart greatly from those in the previous cases. A plausible interpretation of these results is that Eq. (111), although not necessary when $\sigma_n = 0$, is still satisfied for reasons of continuity.

Equation (111) expresses the vanishing of the curl of the spheroidal displacement.

Numerical examples seem to indicate that neutral stratification within the liquid core is not mandatory. However, if it does happen to exist, the deformations corresponding to $\sigma = 0$ are not arbitrary; rather they are determinate.

Elimination of the arbitrary parameter E between the R- and H-equations of the set (110) will give rise to

$$\frac{H}{R} = \frac{H^*}{R^*} - \frac{4\pi G}{\gamma_0} \frac{\lambda X}{R}. \tag{112}$$

When evaluated at $r = b$, this relation represents an additional condition that is valid at the core–mantle interface. Attention should be given to the fact that within the H^*-factor, $\rho_0(b)$ must be interpreted as the value of the density existing at the surface of the liquid core, whereas within the H-factor, $\rho_0(b)$ should signify the value of the density at the bottom of the mantle.

Equations (61) and (112) and the vanishing of the tangential stress

$$F = \mu\left(V' - \frac{V}{r} + \frac{U}{r}\right) \tag{113}$$

at $r = b$ represent three boundary conditions that are valid at the bottom of the mantle.

To ascertain the elastic deformations across the mantle, one should solve Eqs. (53)–(55) with $\sigma = 0$, starting with the above-mentioned three boundary conditions at the bottom of the mantle, with the purpose of achieving the other three boundary conditions that are valid at the free surface.

4.06 NUMERICAL RESULTS

We plan to provide some information on the numerical work that has been undertaken by various authors for the purpose of ascertaining the free and forced oscillations of an elastic Earth, including its elastic deformations conceived as the limit of very long period oscillations.

Love's (1911) original calculations of the free spheroidal oscillations of Earth gave a period of 60 min. They were based on a homogeneous, solid model having a mean rigidity of $\bar{\mu} = 8.9 \cdot 10^{11}$ dyn/cm^2 and $\bar{\lambda} = \bar{\mu}$.

On the other hand, a homogeneous model of the solid Earth with properties averaging Bullen's model B over the whole volume would give rise to a value of $\bar{\rho}_0 = 5.517$ g/cm^3 for the mean density, and $\bar{\mu} = 1.463 \cdot 10^{12}$ dyn/cm^2 and $\bar{\lambda} = 2.402\bar{\mu}$ for the Lamé parameters; simple calculations would reveal its period of spheroidal oscillations to be 44.3 min.

The Kamchatka earthquake of 1952 gave evidence of the existence of two fundamental periods: 57 and 100 min; they were interpreted as pertaining to free oscillations of the spheroidal type.

In a pioneering paper, Alterman *et al.* (1959) numerically solved on a computer, using the Runge–Kutta method, both the spheroidal and toroidal oscillation equations for a nonrotating Earth. These were their results:

1. The periods of the spheroidal oscillations for a Bullen B Earth model corresponding to the spherical harmonics of order $n = 0, 2, 3, 4$, were found to be

n	Period (min)
0	20.0
2	53.7
3	35.5
4	25.7

2. For the same model, the toroidal oscillations were found to have the following periods:

n	Period (min)
2	44.1
3	28.6
4	21.9

To satisfy all the boundary conditions that must be valid at the various interfaces, one must choose values for the frequency σ to be the roots of a determinant equation $D(\sigma) = 0$ whose order depends on the number of conditions. Following this procedure for the case of the Bullen B model and for $n = 2$, the above-mentioned authors found six roots that gave rise to the following periods: 8.0, 9.8, 15.5, 24.8, 53.7 and 100.9 min. The last period happens to be in good agreement with the 100-min period of the Kamchatka earthquake. The same authors also found that the oscillations having the period of 100.9 min exist only within the core, because their amplitudes in the mantle are quite negligible. The Bullen B model was found to have also core oscillations for $n = 3$ and 4 with periods of 85.5 and 76.5 min, respectively.

The bodily tides of the solid Earth induced by a periodic tidal potential of frequency σ were also determined by the same authors for the Bullen B model, giving rise to the following values for the Love numbers h_2, k_2, l_2:

τ (hr)	h_2	k_2	l_2
6	0.629	0.309	0.0888
12	0.618	0.304	0.0883
∞	0.590	0.275	0.0857

h_2 exhibits a variation of 7%, k_2 of 13%, and l_2 of 4% in going from the $\tau = 6$ to the $\tau = \infty$ value. Pekeris and Accad (1972) have given a number of numerical solutions pertaining to the core, using the M_3 Earth model. On the one hand, they have numerically solved the equations of motion for $\mu = 0$; on the other hand, they have used the asymptotic theory, set forth in Section 4.02; the results of the two approaches seem to be in good agreement.

The M_3 model was originally developed by Landisman *et al.* (1965). It was subsequently modified by Pekeris (1966); the density within the mantle was left unchanged, whereas the density within the core was changed in such a manner as to leave the value of β constant throughout the core; three values of β were chosen: -0.2, 0, and 0.2. This result was achieved by the simultaneous solution of the differential system relating the variation of density with gravity in the core:

$$\rho'_0 = -(\rho_0/v_p^2)(1-\beta)\gamma_0, \qquad \gamma'_0 = -(2/r)\gamma_0 + 4\pi G\rho_0. \tag{114}$$

Here v_p is the known velocity of the P-waves; the solutions to the above system must be constrained by the value of Earth's total mass and of its moment of inertia.

Within a core exhibiting unstable stratification (for which $\beta > 0$), the stress λX was found to tend to zero as $\sigma \to 0$ except for a thin boundary layer near the core surface where the stress remains finite. In the limit, when $\sigma = 0$, the stress is discontinuous at the core surface. The radial displacement U also suffers a discontinuity in going from a finite value within the core to a different finite value in the boundary layer. Within this boundary layer, the variables can be represented by exponentials falling off from the core surface. The Love numbers were found to decrease monotonically with increasing period. No free oscillations of the core were found having periods larger than 53.7 min.

For a stably stratified core ($\beta < 0$), an infinite number of free oscillations were found for both the stress and the radial displacement with periods larger than 53.7 min; their amplitudes are confined to the core and are nonexisting in the mantle. The corresponding Love numbers also have a tendency to decrease with increasing period; however, they display resonance spikes corresponding to the periods of free oscillations.

Table 4.1, adapted from Pekeris and Accad (1972), shows the values of h_2, k_2, l_2 for the three stratifications of the M_3 model core corresponding to various periods of the forcing function.

Varga (1974) performed a number of comparative studies on the Love numbers for a nonloaded surface using various Earth models: Bullen's A and B models, Molodenskii's model, and Bullard's model A.

Table 4.1

Values of the Love Numbers h_2, k_2, l_2 for the M_3 Earth Model with Uniform Polytropic Cores with Stratification Parameter $\beta = -0.2, 0, 0.2$[a]

Period (hr)	h_2			k_2			l_2		
	−0.2	0	0.2	−0.2	0	0.2	−0.2	0	0.2
6	0.6243	0.6253	0.6256	0.3068	0.3076	0.3084	0.0847	0.0847	0.0848
12	0.6162	0.6152	0.6154	0.3022	0.3028	0.3035	0.0840	0.0843	0.0843
24	0.6120	0.6127	0.6128	0.3009	0.3016	0.3023	0.0841	0.0842	0.0842
∞[b]	0.6118	0.6119	0.6119	0.3005	0.3012	0.3019	0.0840	0.0841	0.0842

[a] Results adapted from Pekeris and Accad (1972).
[b] Derived from asymptotic solution.

By varying certain parameters, he was able to provide the following results:

A. The value $\rho_0(a)$ of the density at the outermost surface and the value $\rho_0(b)$ of the density at the core–mantle interface have very slight effects on the values of the Love numbers at $r = a$.

B. The ellipticity of the core surface, which is provided by the solution $\eta_2(b)$ of the Radau equation (see Chapter II), has more influence on the values of the Love numbers; thus, in going from $\eta_2(b) = 0$ to $\eta_2(b) = 0.0855$, he found that h_2 varies from 0.616 to 0.623, k_2 from 0.304 to 0.312, and l_2 from 0.088 to 0.084.

C. Changes due to the values of $\mu(b)$ are given in Table 4.2.

Table 4.2

Dependence of the Love Numbers on the Rigidity of the Core[a]

$\mu(b) \cdot 10^{12}$ (dyn/cm^2)	h_2	k_2	l_2	h_3	k_3	l_3
0	0.623	0.312	0.088	0.312	0.101	0.014
0.00035	0.621	0.309	0.088	0.307	0.101	0.014
0.00345	0.617	0.305	0.088	0.307	0.101	0.014
0.01726	0.610	0.301	0.088	0.305	0.101	0.014
0.03452	0.605	0.298	0.088	0.303	0.100	0.014
0.1726	0.564	0.279	0.088	0.288	0.097	0.017
0.3452	0.522	0.258	0.088	0.276	0.092	0.019
1.0356	0.412	0.204	0.088	0.234	0.079	0.024
1.7260	0.349	0.173	0.088	0.222	0.068	0.024
2.7616	0.293	0.145	0.087	0.192	0.059	0.026

[a] Results are due to Varga (1974).

D. Variations due to the relative radius $s_0 = b/a$ of the core are given in Table 4.3.

Table 4.3

Dependence of the Love Numbers on the Radius of the Core[a]

b/a	h_2	k_2	l_2	h_3	k_3	l_3
0.555	0.639	0.321	0.082	0.320	0.106	0.014
0.545	0.618	0.306	0.085	0.305	0.101	0.015
0.535	0.595	0.292	0.083	0.292	0.096	0.016

[a] Results are due to Varga (1974).

Both Longman (1962) and Farrell (1972) have given a number of results on the load numbers (h', k', l') for a nonrotating Earth using the Gutenberg model. A detailed description of this model has been given by Alterman *et al.* (1961). Longman has given the values of the load numbers of orders 0–40. Farrell has gone beyond that; he was the first to use the asymptotic theory of the Boussinesq flat plate. Results by these two authors, which were obtained by numerical integration of the equations of motion, are given in Table 4.4.

Farrell has also provided numerical results pertaining to the spheroidal deformations of a loaded Earth as a function of the angular distance from the loading line. He pioneered the procedure described in Section 4.04 for obtaining the spheroidal deformations by finding the sum of a infinite series of Legendre polynomials.

Table 4.5 is adapted from Farrell's results; it provides radial and tangential displacements, gravity variation, tilt, and the $\epsilon_{\theta\theta}$-component of the strain tensor due to surface loading.

More recently, Wahr (1981) has numerically integrated the equations of motion pertaining to a rotating, slightly spheroidal Earth. He used a mathematical model developed by Smith (1974) which is akin to the analytic model of a rotating Earth described in Chapter III.

Wahr solved the problem of the forced oscillations generated by the various tides which can be identified by their Doodson numbers (see Chapter VI); he primarily used a variant of the 1066A Earth model in which the density distribution in the core was modified in order to reduce its stratification to neutral stability. This can be achieved by using Eqs. (114) with $\beta = 0$; it implies also the vanishing of the Brunt–Vaisala frequency N^2, since the two parameters are related by Eq. (86) in Chapter III.

Table 4.4

Load Numbers for Gutenberg Earth Model According to Longman and Farrell[a]

n	$-h'_n$		$-nk'_n$		nl'_n	
	Longman	Farrell	Longman	Farrell	Longman	Farrell
2	1.007	1.001	0.620	0.615	0.060	0.059
3	1.059	1.052	0.591	0.585	0.225	0.223
4	1.059	1.053	0.532	0.528	0.248	0.247
5	1.093	1.088	0.520	0.516	0.245	0.243
6	1.152	1.147	0.540	0.535	0.246	0.245
7	1.223		0.574		0.259	
8	1.296	1.291	0.608	0.604	0.272	0.269
9	1.369		0.648		0.288	
10	1.439	1.433	0.690	0.682	0.300	0.303
12	1.572		0.768		0.336	
14	1.691		0.840		0.378	
16	1.798		0.896		0.416	
18	1.902	1.893	0.972	0.952	0.450	0.452
20	1.994		1.020		0.500	
22	2.078		1.078		0.528	
24	2.156		1.128		0.552	
26	2.223		1.170		0.598	
28	2.291		1.204		0.616	
30	2.351		1.230		0.660	
32	2.408	2.379	1.280	1.240	0.672	0.680
34	2.455		1.292		0.714	
36	2.497		1.332		0.720	
38	2.535		1.330		0.760	
40	2.581		1.360		0.760	
56		2.753		1.402		0.878
100		3.058		1.461		0.973
180		3.474		1.591		1.023
325		4.107		1.928		1.212
550		4.629		2.249		1.460
1,000		4.906		2.431		1.623
1,800		4.953		2.465		1.656
3,000		4.954		2.468		1.657
10,000		4.956		2.469		1.657
∞		5.005		2.482		1.673

[a] From Longman (1962) and Farrell (1972).

Table 4.5

Spheroidal Deformations for the Gutenberg Earth Model[a]

Distance from load θ (deg)	Radial displacement $\times 10^{12}(a\theta)$	Tangential displacement $\times 10^{12}(a\theta)$	Gravity variation $\times 10^{18}(a\theta)$	Tilt $\times 10^{12}(a\theta)^2$	Strain, $\epsilon_{\theta\theta}$ $\times 10^{12}(a\theta)^2$
10^{-4}	−33.64	−11.25	−77.87	33.64	11.248
10^{-3}	−33.56	−11.25	−77.69	33.64	11.248
10^{-2}	−32.75	−11.24	−75.92	33.64	11.253
0.1	−25.41	−10.36	−59.64	32.35	11.866
0.25	−18.36	−8.024	−43.36	25.29	11.083
0.5	−14.91	−6.333	−34.32	17.85	7.538
0.8	−13.69	−6.050	−30.59	16.25	6.019
1.0	−13.01	−5.997	−28.75	16.32	6.170
1.2	−12.31	−5.881	−27.03	16.33	6.535
2.0	−9.757	−4.981	−21.38	14.95	7.114
2.5	−8.519	−4.388	−18.74	13.68	6.830
3.0	−7.533	−3.868	−16.64	12.38	6.332
4.0	−6.131	−3.068	−13.59	10.09	5.261
5.0	−5.237	−2.523	−11.55	8.27	4.297
6.0	−4.660	−2.156	−10.16	6.90	3.445
7.0	−4.272	−1.915	−9.169	5.94	2.765
8.0	−3.999	−1.754	−8.425	5.23	2.230
9.0	−3.798	−1.649	−7.848	4.72	1.800
10.0	−3.640	−1.582	−7.379	4.38	1.485
20.0	−2.619	−1.386	−4.725	3.36	0.712
25	−2.103	−1.312	−3.804	3.31	0.812
30	−1.530	−1.211	−2.951	3.29	1.104
40	−0.292	−0.926	−1.427	2.94	2.040
50	0.848	−0.592	−0.279	1.94	2.928
60	1.676	−0.326	0.379	0.39	3.253
70	2.083	−0.223	0.557	−1.25	2.829
80	2.057	−0.310	0.353	−2.71	1.789
90	1.643	−0.555	−0.110	−3.76	0.395
100	0.920	−0.894	−0.713	−4.31	−1.094
110	−0.025	−1.247	−1.357	−4.39	−2.475
120	−1.112	−1.537	−1.980	−4.18	−3.656
130	−2.261	−1.706	−2.557	−3.72	−4.638
140	−3.405	−1.713	−3.076	−3.12	−5.473
150	−4.476	−1.540	−3.530	−2.44	−6.191
160	−5.414	−1.182	−3.918	−1.67	−6.825
170	−6.161	−0.657	−4.243	−0.83	−7.441
180	−6.663	0	−4.514	0	−8.203

[a] Applied load is 1 kg; units are MKS; $a = 6.371 \cdot 10^6$ m. In the normalization, θ must be expressed in radians. Results are due to Farrell (1972).

Wahr's model assumes a slightly spheroidal Earth represented by the equation

$$r(\theta)/a = 1 - \tfrac{2}{3} e_0 P_2(\cos\theta),$$

where $a = 6371$ km, $e_0 = 0.00334$, and the gravity at the equator is taken as $g(a) = 979.8259$ cm/sec^2. His results depend on the colatitude θ of the location on the spheroid; however, the corresponding coefficients are small, as should be expected.

Let us briefly describe some of the results given by Wahr, limiting ourselves, however, to the case of the semidiurnal tides ($l = m = 2$); for other details pertaining to different kinds of tides, reference should be made to the aforementioned author's publication.

In what follows, we have denoted by Y_l^m the spherical harmonic functions which have been defined by Eq. (64) in Chapter II, and by A the frequency-dependent amplitude of the given tide ($l = m = 2$), which is given in Table 6.1.

The variation in the magnitude of gravity is

$$\frac{g^*(a,\theta)}{g(a)} = -\frac{2}{a} A(1.160 Y_2^2 - 0.005 Y_4^2); \tag{115}$$

the component of the tilt in the ϕ-direction (i.e., east) is

$$t_2^*(\theta) = \frac{A}{a}\left(\frac{1.378}{\sin\theta} Y_2^2 - 0.002 \frac{\partial Y_3^2}{\partial\theta}\right); \tag{116}$$

the north–south component of the strain tensor is

$$\epsilon_{\theta\theta} = \frac{A}{r(\theta)}\left\{0.609 Y_2^2 + [0.085 + 0.001 P_2(\cos\theta)] \frac{\partial^2 Y_2^2}{\partial\theta^2}\right\}; \tag{117}$$

the distortion in the potential is

$$V_1^*(a,\theta) = -(0.302) A g(a)\left[\left(\frac{a}{r}\right)^3 Y_2^2 - 0.003\left(\frac{a}{r}\right)^5 Y_4^2\right]; \tag{118}$$

the radial displacement is

$$u = (0.609) A[1 + 0.001 P_2(\cos\theta)] Y_2^2; \tag{119}$$

and the component of the displacement in the θ-direction is

$$v = (0.0852) A\left\{[1 + 0.014 P_2(\cos\theta)] \frac{\partial Y_2^2}{\partial\theta} - 0.001 \frac{\partial Y_4^2}{\partial\theta} + \frac{0.002}{\sin\theta} Y_3^2\right\}. \tag{120}$$

Thus we can write, e.g.,

$$g^*|g = -\frac{2A}{a}\left[1.160 - (0.005)\frac{\sqrt{3}}{2}(7\cos^2\theta - 1)\right]Y_2^2, \tag{121}$$

which shows the dependence on colatitude of the gravity factor within brackets; this factor will be 1.164 at the equator ($\theta = \pi/2$) but only 1.149 at 45° latitude.

Chapter V Precessional and Nutational Motions of a Rigid Earth

5.00 INTRODUCTION

In order to reach a better understanding of the role that geophysical perturbations play on the rotational motion of Earth, we shall devote this chapter to a study of the precessional and nutational motion of a rigid, massive body.

We start by considering the Euler equations for the motion of the body axes of a rigid Earth in inertial space and establish the Poisson equations as their fundamental approximations.

We proceed by investigating first the Eulerian motion for an unperturbed Earth and then by examining the perturbing effects of the lunisolar forces. We also analyze the Poinsot interpretation of the motion by means of rolling cones.

We next follow Woolard's approach in reaching the numerical solutions of the Poisson and other related equations by making use of Brown's lunar theory and Newcomb's results on solar perturbations. We provide a shortened table of numerical results adapted from Woolard so that the reader may get a clear understanding of the magnitude of the numerical computation effort.

The problem of updating the numerical values for the precession and nutation of Earth has required that we examine the motion of the ecliptic and equinox of date with respect to the ecliptic of epoch.

We complete this chapter by considering the dynamical behavior of a binary system of rigid bodies of comparable size. Using the Lagrangian formulation, we reach the fundamental equations of motion that reflect the influence of the relative motion of the centroids upon the rotational motion of the body axes. We finally mention some open questions that must be solved to arrive at a workable model of the dynamics of a binary system.

5.01 MOTION OF A RIGID EARTH ABOUT ITS CENTER OF GRAVITY

We visualize Earth as a rigid body; we denote by O its center of mass and by ξ, η, ζ the orthogonal set of its principal axes of inertia. Let A, B, C, with $A < B < C$, be the principal moments of inertia. The ζ-axis about which the moment of inertia is a maximum is referred to as the axis of figure, and the intersection of the Earth with the plane through O perpendicular to ζ is called the equator of figure.

We also consider an inertial orthogonal frame $(O; x, y, z)$ with the same origin O; the (xy)-plane coincides with the fixed ecliptic, and the positive direction of the x-axis is pointing toward the fixed mean equinox of an adopted epoch. We introduce the three Euler angles: α, the precession angle; β, the rotation angle; and γ, the nutation angle. The inertial frame $(O; x, y, z)$ can be taken into the rotating frame $(O; \xi, \eta, \zeta)$ of the body axes by means of three rotations in the positive (i.e., counterclockwise) direction having magnitudes α, β, γ. Let us denote by l the lines of nodes, i.e., the intersection of the equator of figure [$(\xi\eta)$-plane] with the fixed ecliptic [(xy)-plane].

A first rotation through α about the fixed z-axis takes the x-axis into the line of nodes and the y-axis into the new position y_1; this rotation is represented by

$$\begin{bmatrix} x \\ y \\ z \end{bmatrix} = \begin{bmatrix} \cos\alpha & -\sin\alpha & 0 \\ \sin\alpha & \cos\alpha & 0 \\ 0 & 0 & 1 \end{bmatrix} \begin{bmatrix} l \\ y_1 \\ z \end{bmatrix}.$$

A second rotation of amplitude γ about the line of nodes l will take the z-axis into the ζ-axis and y_1-axis into the new position y_2; its representation is

$$\begin{bmatrix} l \\ y_1 \\ z \end{bmatrix} = \begin{bmatrix} 1 & 0 & 0 \\ 0 & \cos\gamma & -\sin\gamma \\ 0 & \sin\gamma & \cos\gamma \end{bmatrix} \begin{bmatrix} l \\ y_2 \\ \zeta \end{bmatrix}.$$

A third rotation of amplitude β about the ζ-axis takes l into ξ and y_2 into η; it is represented by

$$\begin{bmatrix} l \\ y_2 \\ \zeta \end{bmatrix} = \begin{bmatrix} \cos\beta & -\sin\beta & 0 \\ \sin\beta & \cos\beta & 0 \\ 0 & 0 & 1 \end{bmatrix} \begin{bmatrix} \xi \\ \eta \\ \zeta \end{bmatrix}.$$

It follows that the transformation from the inertial set $(O; x, y, z)$ to the body axes set $(O; \xi, \eta, \zeta)$ can be represented by

$$\begin{bmatrix} x \\ y \\ z \end{bmatrix} = A \begin{bmatrix} \xi \\ \eta \\ \zeta \end{bmatrix}, \tag{1}$$

where

$$A = \begin{bmatrix} \cos\alpha\cos\beta - \sin\alpha\sin\beta\cos\gamma & -\cos\alpha\sin\beta - \sin\alpha\cos\beta\cos\gamma & \sin\alpha\sin\gamma \\ \sin\alpha\cos\beta + \cos\alpha\sin\beta\cos\gamma & -\sin\alpha\sin\beta + \cos\alpha\cos\beta\cos\gamma & -\cos\alpha\sin\gamma \\ \sin\beta\sin\gamma & \cos\beta\sin\gamma & \cos\gamma \end{bmatrix} \tag{2}$$

is the product of the previous three matrices.

Figure 5.1 shows the relative position of the three Euler angles: $(xl) = \alpha$, $(z\zeta) = \gamma$, and $(l\xi) = \beta$.

The motion of the Earth about its centroid O can be visualized as the rotational motion of its principal axes of inertia (ξ, η, ζ) due to the rotational vector $\boldsymbol{\omega}$ passing through the centroid O and admitting of the following decomposition:

$$\boldsymbol{\omega} = \dot{\alpha}\mathbf{k}_1 + \dot{\beta}\mathbf{k} + \dot{\gamma}\mathbf{n}.$$

Here dots denote derivatives with respect to time, $\mathbf{k}_1$ and $\mathbf{k}$ are the unit vectors along the z- and ζ-axes, respectively, and $\mathbf{n}$ is the unit vector along the line of nodes.

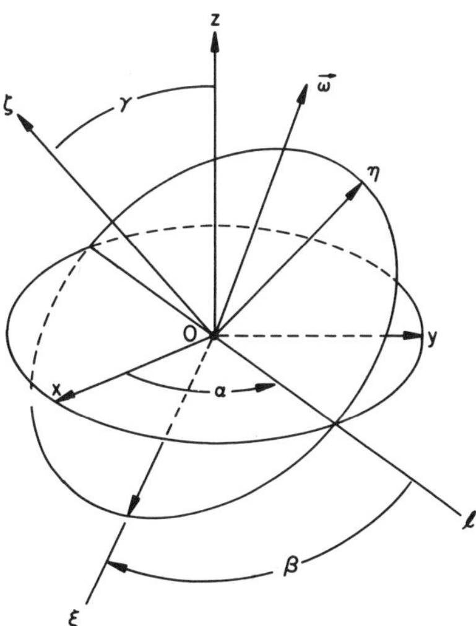

Fig. 5.1 Definition of the Euler angles.

The components $\omega_1, \omega_2, \omega_3$ of $\boldsymbol{\omega}$ along the body axes (ξ, η, ζ) are then given by

$$
\begin{aligned}
\omega_1 &= \dot{\gamma}\cos\beta + \dot{\alpha}a_{31} = \dot{\gamma}\cos\beta + \dot{\alpha}\sin\beta\sin\gamma, \\
\omega_2 &= -\dot{\gamma}\sin\beta + \dot{\alpha}a_{32} = -\dot{\gamma}\sin\beta + \dot{\alpha}\cos\beta\sin\gamma, \\
\omega_3 &= \dot{\beta} + \dot{\alpha}\cos\gamma,
\end{aligned}
\tag{3}
$$

where a_{31} and a_{32} are elements of the above matrix A of direction cosines.

Equations (3) are Euler's kinematic equations; they can be solved for $\dot{\alpha}$, $\dot{\beta}$, $\dot{\gamma}$ in terms of $\omega_1, \omega_2, \omega_3$ as follows:

$$
\begin{aligned}
\dot{\alpha}\sin\gamma &= \omega_1\sin\beta + \omega_2\cos\beta, \\
\dot{\gamma} &= \omega_1\cos\beta - \omega_2\sin\beta, \\
\dot{\beta} &= \omega_3 - (\cos\gamma/\sin\gamma)(\omega_1\sin\beta + \omega_2\cos\beta).
\end{aligned}
\tag{4}
$$

To obtain the equations of the rotational motion, we shall use the Lagrangian formulation, according to which

$$
\frac{d}{dt}\left(\frac{\partial L}{\partial \dot{q}_i}\right) = \frac{\partial L}{\partial q_i}, \qquad i = 1, 2, 3,
$$

where q stands for any one of the angles α, β, γ and the Lagrangian is

$$
L = U(q) + T(q, \dot{q}) = U(\alpha, \beta, \gamma) + \tfrac{1}{2}(A\omega_1^2 + B\omega_2^2 + C\omega_3^2), \tag{5}
$$

where U is the force function for the gravitational attraction of external bodies. Solving for the derivatives of the components of the rotational velocity, we obtain

$$
\begin{aligned}
A\frac{d\omega_1}{dt} - (B - C)\omega_2\omega_3 &= \frac{\partial U}{\partial \gamma}\cos\beta + \frac{\sin\beta}{\sin\gamma}\left(\frac{\partial U}{\partial \alpha} - \cos\gamma\frac{\partial U}{\partial \beta}\right), \\
B\frac{d\omega_2}{dt} - (C - A)\omega_3\omega_1 &= -\frac{\partial U}{\partial \gamma}\sin\beta + \frac{\cos\beta}{\sin\gamma}\left(\frac{\partial U}{\partial \alpha} - \cos\gamma\frac{\partial U}{\partial \beta}\right), \\
C\frac{d\omega_3}{dt} - (A - B)\omega_1\omega_2 &= \frac{\partial U}{\partial \beta}.
\end{aligned}
\tag{6}
$$

These are the dynamical equations of the rotational motion about O. For Earth, we can take as first-order approximation $A = B$; in this case, because of symmetry, we must have $\partial U/\partial \beta = 0$. The above equations reduce then to

$$
\begin{aligned}
\frac{d\omega_1}{dt} - \frac{A - C}{A}\omega_2\omega_3 &= \frac{1}{A}\left(\frac{\partial U}{\partial \gamma}\cos\beta + \frac{\sin\beta}{\sin\gamma}\frac{\partial U}{\partial \alpha}\right), \\
\frac{d\omega_2}{dt} - \frac{C - A}{A}\omega_3\omega_1 &= \frac{1}{A}\left(-\frac{\partial U}{\partial \gamma}\sin\beta + \frac{\cos\beta}{\sin\gamma}\frac{\partial U}{\partial \alpha}\right), \\
\omega_3 &= \text{const.}
\end{aligned}
\tag{7}
$$

These equations should be solved simultaneously with Eqs. (4) in order to determine the set (α, β, γ) relating the two frames and $(\omega_1, \omega_2, \omega_3)$ establishing the rotational attitude of the Earth.

One can also write a differential system pertaining to α, β, γ alone and bypass the $\omega_1, \omega_2, \omega_3$. This can be achieved by time differentiation of Eqs. (4) and by eliminating ω_1, ω_2, ω_3 and their derivatives through the use of Eqs. (3) and (7). The same result can be obtained more expeditiously by rewriting the expression of T for the kinetic energy of the system given by Eq. (5) in terms of $\dot{\alpha}$, $\dot{\beta}$, $\dot{\gamma}$ and by means of Eqs. (3). For the case $A = B$, we get

$$2T = A(\dot{\alpha}^2 \sin^2 \gamma + \dot{\gamma}^2) + C(\dot{\beta} + \dot{\alpha} \cos \gamma)^2. \tag{8}$$

Through the Lagrangian formulation and the knowledge that $\omega_3 = \text{const.}$, we obtain the following system:

$$\begin{aligned}
\dot{\gamma} &= -\frac{1}{C\omega_3 \sin \gamma} \frac{\partial U}{\partial \alpha} + \frac{A}{C\omega_3} \left[\frac{d}{dt} (\dot{\alpha} \sin \gamma) + \dot{\alpha}\dot{\gamma} \cos \gamma \right], \\
\dot{\alpha} &= \frac{1}{C\omega_3 \sin \gamma} \frac{\partial U}{\partial \gamma} + \frac{A}{C\omega_3} \left(\dot{\alpha}^2 \cos \gamma - \frac{\ddot{\gamma}}{\sin \gamma} \right), \\
\dot{\beta} &= \omega_3 - \dot{\alpha} \cos \gamma.
\end{aligned} \tag{9}$$

Except for the approximation $A = B$, the above system constitutes the most general set of equations for studying the motion of the body axes with respect to the inertial frame.

The last two terms appearing in the right-hand sides of the $\dot{\alpha}$ and $\dot{\gamma}$ equations turn out to be so much smaller than the corresponding first terms that we can neglect them as a first-order approximation. In doing so we obtain the Poisson equations:

$$\dot{\gamma} = -\frac{1}{C\omega_3 \sin \gamma} \frac{\partial U}{\partial \alpha}, \qquad \dot{\alpha} = \frac{1}{C\omega_3 \sin \gamma} \frac{\partial U}{\partial \gamma}. \tag{10}$$

5.02 PRECESSION AND NUTATION OF THE ROTATIONAL AXIS

Let us now consider the instantaneous rotational velocity vector $\boldsymbol{\omega}$. We shall denote by θ_r the angle that this vector makes with the ζ-axis (that is to say, its colatitude), and by ϕ_r the longitude of its projection $\overline{OJ}$ onto the $(\xi\eta)$-plane reckoned from the ξ-axis [angle (ξJ) in Fig. 5.2]. The position of $\boldsymbol{\omega}$ with respect to the inertial frame can be defined in terms of the inclination angle ϵ_1 that the equator of rotation makes with the fixed ecliptic and the longitude α_r of its node onto the fixed ecliptic reckoned from the x-axis. If

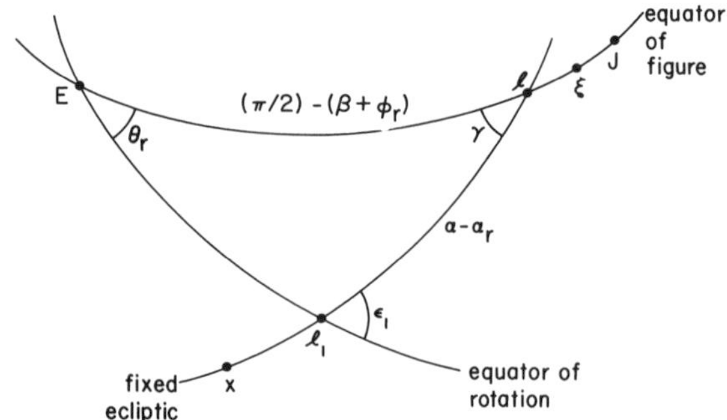

Fig. 5.2 Position of the equator of rotation and the equator of figure with respect to the fixed ecliptic.

we consider the three great circles—equator of figure, equator of rotation, and fixed ecliptic—we have the spherical triangle (El_1l) of Fig. 5.2, where $(xl) = \alpha$, $(xl_1) = \alpha_r$, $(l\xi) = \beta$, $(\xi J) = \phi_r$, $(EJ) = \pi/2$, $(z\zeta) = \gamma$, and $(\zeta\omega) = \theta_r$. Solving this triangle for ϵ_1 and $\alpha - \alpha_r$, we have

$$\begin{aligned} \cos\epsilon_1 &= \cos\theta_r \cos\gamma - \sin\theta_r \sin\gamma \sin(\beta + \phi_r), \\ \sin\epsilon_1 \sin(\alpha - \alpha_r) &= \sin\theta_r \cos(\beta + \phi_r), \\ \sin\epsilon_1 \cos(\alpha - \alpha_r) &= \cos\theta_r \sin\gamma + \sin\theta_r \cos\gamma \sin(\beta + \phi_r). \end{aligned} \tag{11}$$

In effect, the vector $\boldsymbol{\omega}$ lies very close to the axis of figure (ζ-axis), so that θ_r, $\gamma - \epsilon_1$, and $\alpha - \alpha_r$ are small quantities. We can therefore apply the small angle approximation to the three equations of the set (11) to obtain

$$\begin{aligned} \alpha - \alpha_r &= \frac{\theta_r}{\sin\gamma} \cos(\beta + \phi_r), \\ \gamma - \epsilon_1 &= -\theta_r \sin(\beta + \phi_r). \end{aligned} \tag{12}$$

As the next step we wish to express the time variation of both ϵ_1 and α_r. For this purpose we introduce the unit vector $\mathbf{u}$ along the instantaneous rotational axis. We then relate the components of this vector with respect to the axes of the inertial frame, which we can write as

$$\mathbf{u} = \begin{bmatrix} \sin\epsilon_1 \sin\alpha_r \\ -\sin\epsilon_1 \cos\alpha_r \\ \cos\epsilon_1 \end{bmatrix},$$

to the components of the same vector with respect to the body axes; this relation can be achieved by means of the matrix A of direction cosines between the two sets of axes. We get

$$A\begin{bmatrix}\omega_1\\ \omega_2\\ \omega_3\end{bmatrix} = \omega\mathbf{u}.$$

Time differentiation of the above equation with respect to the rotating frame will yield

$$A\begin{bmatrix}\dot{\omega}_1\\ \dot{\omega}_2\\ 0\end{bmatrix} + \frac{\partial A}{\partial t}\begin{bmatrix}\omega_1\\ \omega_2\\ \omega_3\end{bmatrix} = \dot{\omega}\mathbf{u} + \omega\dot{\mathbf{u}}. \tag{13}$$

The three rows that constitute the matrix A are the components with respect to the rotating system of the unit vectors along the reference axes of the inertial system. We shall denote these row vectors by $\mathbf{v}_i$ $(i = 1, 2, 3)$. The time derivatives of these vectors in the inertial frame shall vanish, and when we relate these derivatives to the rotating frame we get

$$\frac{d\mathbf{v}_i}{dt} = 0 = \frac{\partial \mathbf{v}_i}{\partial t} + \mathbf{v}_i \times \boldsymbol{\omega},$$

where we have used the symbol of partial derivative to denote differentiation in the rotating system. It follows that

$$\frac{\partial \mathbf{v}_i}{\partial t} = \boldsymbol{\omega} \times \mathbf{v}_i$$

and that

$$\boldsymbol{\omega} \cdot \frac{\partial \mathbf{v}_i}{\partial t} = \boldsymbol{\omega} \cdot (\boldsymbol{\omega} \times \mathbf{v}_i) = 0.$$

This is equivalent to saying that the second term on the left-hand side of Eq. (13) vanishes.

Solving Eq. (13) for the derivatives of ϵ_1 and α_r, we get

$$\begin{aligned}-\omega\dot{\epsilon}_1 &= [\cos(\alpha - \alpha_r)\cos\gamma\cos\epsilon_1 + \sin\gamma\sin\epsilon_1](\dot{\omega}_1\sin\beta + \dot{\omega}_2\cos\beta)\\ &\quad + \sin(\alpha - \alpha_r)\cos\epsilon_1(\dot{\omega}_1\cos\beta - \dot{\omega}_2\sin\beta),\\ (\omega\sin\epsilon_1)\dot{\alpha}_r &= \cos(\alpha - \alpha_r)(\dot{\omega}_1\cos\beta - \dot{\omega}_2\sin\beta)\\ &\quad - (\cos\gamma)\sin(\alpha - \alpha_r)(\dot{\omega}_1\sin\beta + \dot{\omega}_2\cos\beta).\end{aligned} \tag{14}$$

These are exact equations in which all terms are accounted for, the only simplifying assumption being the equality $A = B$ of the two moments of inertia.

To simplify the previous equations, we first proceed to eliminate the terms in $\dot{\omega}_1$ and $\dot{\omega}_2$ by making use of Eqs. (7), (3), and (9). This operation yields

$$\begin{aligned}\dot{\omega}_1 \sin\beta + \dot{\omega}_2 \cos\beta &= \frac{1}{C\sin\gamma}\frac{\partial U}{\partial\alpha} + \frac{C-A}{C}\left[\dot{\alpha}\dot{\gamma}\cos\gamma + \frac{d}{dt}(\dot{\alpha}\sin\gamma)\right],\\ \dot{\omega}_1\cos\beta - \dot{\omega}_2\sin\beta &= \frac{1}{C}\frac{\partial U}{\partial\gamma} + \frac{A-C}{C}(\dot{\alpha}^2\sin\gamma\cos\gamma - \ddot{\gamma}).\end{aligned} \tag{15}$$

We next note that the leading terms in these two expressions are those containing the partial derivatives of the force function U since the following terms are quadratic terms and furthermore are multiplied by the small factor $(C - A)/C$; the Poisson equations (10) can then be used to represent these leading terms by

$$-\omega_3\dot{\gamma}, \qquad \omega_3\dot{\alpha}\sin\gamma. \tag{16}$$

We are now in a position to simplify Eq. (14) by first writing the small angle approximation pertaining to $\alpha - \alpha_r$ and $\gamma - \epsilon_1$ and then substituting either Eq. (15) or (16) within it according to the required approximation. We reach the following results:

$$\begin{aligned}-\omega_3\dot{\epsilon}_1 &= \frac{1}{C\sin\gamma}\frac{\partial U}{\partial\alpha} + \frac{C-A}{C}\left[\dot{\alpha}\dot{\gamma}\cos\gamma + \frac{d}{dt}(\dot{\alpha}\sin\gamma)\right]\\ &\quad + (\alpha - \alpha_r)\cos\gamma\sin\gamma\,\omega_3\dot{\alpha},\\ \omega_3\dot{\alpha}_r\sin\gamma &= \frac{1}{C}\frac{\partial U}{\partial\gamma} + \frac{A-C}{C}(\dot{\alpha}^2\sin\gamma\cos\gamma - \ddot{\gamma})\\ &\quad + (\alpha - \alpha_r)\cos\gamma\,\omega_3\dot{\gamma}.\end{aligned} \tag{17}$$

These equations are equivalent to Eqs. (9) but have the advantage that the second-order terms beyond the Poisson approximation are even smaller than the corresponding terms appearing in Eqs. (9).

We can finally evaluate the time variation of the magnitude ω of the velocity vector $\boldsymbol{\omega}$; let us remark that ω is not a constant, whereas the component ω_3 about the axis of figure is. Differentiating $\omega^2 = \omega_1^2 + \omega_2^2 + \omega_3^2$ with respect to time and using Eqs. (7) and the fact that

$$\omega_1/\omega = \sin\theta_r\cos\phi_r, \qquad \omega_2/\omega = \sin\theta_r\sin\phi_r, \tag{18}$$

we obtain

$$\dot{\omega} = \frac{\sin\theta_r}{A}\left[\frac{\partial U}{\partial\gamma}\cos(\beta + \phi_r) + \frac{1}{\sin\gamma}\frac{\partial U}{\partial\alpha}\sin(\beta + \phi_r)\right]. \tag{19}$$

5.03 EULERIAN MOTION AND LUNISOLAR PERTURBATIONS

The solutions of Eqs. (7) will determine the components ω_1, ω_2 of $\boldsymbol{\omega}$, that is to say, the motion of the instantaneous rotational axis with respect to the body axes. The solutions of Eqs. (4) or (9), on the other hand, establish the location of the body axes with respect to the inertial frame.

Let us consider Eqs. (7) and solve them by the variation of parameters method. This amounts to solving first the simplified system obtained by neglecting the right-hand-side second-order terms due to the force function. We are left with a linear system that describes the Eulerian motion of Earth unperturbed by the lunisolar forces.

If the rotational axis coincides initially with the axis of figure, then $\omega_1 = \omega_2 = 0$, $\omega_3 =$ const. is a solution to the system and the rigid body continues its constant rotation about the axis of figure. In this case, the dynamical equations (4) degenerate to $\alpha =$ const., $\gamma =$ const., and $\dot{\beta} = \omega$. Thus the axis of rotation remains also fixed in space.

Suppose next that the rotational velocity vector $\boldsymbol{\omega}_0$ is originally displaced from the axis of figure by the angle $\theta_{\mathrm{r},0}$; if we introduce the constant

$$\mu = \frac{C - A}{A}\,\omega_3, \tag{20}$$

we get as solutions of Eq. (7)

$$\begin{aligned} \omega_1 &= f_0 \cos \mu t + g_0 \sin \mu t = h_0 \cos(\mu t - \sigma_0), \\ \omega_2 &= f_0 \sin \mu t - g_0 \cos \mu t = h_0 \sin(\mu t - \sigma_0), \end{aligned} \tag{21}$$

where f_0, g_0 are two arbitrary constants of integration. It follows that

$$f_0 = h_0 \cos \sigma_0, \qquad g_0 = h_0 \sin \sigma_0, \tag{22}$$

so that h_0, σ_0 are also constants. $\boldsymbol{\omega}_0$ has then a constant magnitude

$$\omega_0 = (\omega_3^2 + h_0^2)^{1/2},$$

makes a constant angle $\theta_{\mathrm{r},0}$ with the ζ-axis, with $\cos \theta_{\mathrm{r},0} = \omega_3/\omega_0$, and its projection h_0 onto the $(\xi\eta)$-plane makes the angle $\phi_{\mathrm{r},0} = \mu t - \sigma_0$ with the ξ-axis. $\boldsymbol{\omega}_0$ describes a conical surface about the ζ-axis at a constant rate.

Thus the general solution of the Eulerian motion is a constant angular rotation about an axis which does not coincide with the axis of figure, but which precesses about it at a constant rate. Any perturbing force will disturb this state of affairs, causing the rotational axis to deviate from the axis of figure, or if originally displaced from the axis of figure by the angle $\theta_{\mathrm{r},0}$, these perturbing forces will produce an additional deviation given by the angle $\theta_{\mathrm{r},1}$. In the case of Earth, both the lunisolar perturbations and other causes of a geophysical nature will cause the rotational axis to differ from the axis of figure. The solution to the complete system (7) will be of the form

$$\begin{aligned} \omega_1 &= f\cos\mu t + g\sin\mu t = h\cos(\mu t - \sigma), \\ \omega_2 &= f\sin\mu t - g\cos\mu t = h\sin(\mu t - \sigma), \end{aligned} \tag{23}$$

where $f = h\cos\sigma$, $g = h\sin\sigma$, and

$$\begin{aligned} f &= f_0 + \frac{1}{A}\int\left[\cos(\beta + \mu t)\frac{\partial U}{\partial \gamma} + \frac{\sin(\beta + \mu t)}{\sin\gamma}\frac{\partial U}{\partial \alpha}\right]dt, \\ g &= g_0 + \frac{1}{A}\int\left[\sin(\beta + \mu t)\frac{\partial U}{\partial \gamma} - \frac{\cos(\beta + \mu t)}{\sin\gamma}\frac{\partial U}{\partial \alpha}\right]dt. \end{aligned} \tag{24}$$

Substituting Eqs. (23) into Eq. (4), we get

$$\begin{aligned} \dot\alpha\sin\gamma &= h\sin(\beta + \mu t - \sigma), \\ \dot\gamma &= h\cos(\beta + \mu t - \sigma), \\ \frac{d}{dt}(\beta + \mu t) &= \frac{C}{A}\omega_3 - \dot\alpha\cos\gamma. \end{aligned} \tag{25}$$

Let us note that although in Eq. (24) the unknowns ω_1, ω_2, α, β, γ also appear under the integral sign, they will be second-order terms; we can therefore perform those required integrations by successive steps. We thus obtain the general solution by quadratures depending on six constants of integration: ω_3, f_0, g_0 and the initial values of α, β, γ. These constants must be ascertained by observation.

In the general case, the rotational rate ω is not constant, and the precessional motion of the rotational axis is not uniform because in the expression for the angle $\phi_r = \mu t - \sigma$, σ is also a function of time.

The motion of the instantaneous axis of rotation within the Earth produces variations in geographical latitudes and longitudes; being very small, these variations were not detected until the end of last century. The motion of the rotational axis in space, on the contrary, is a slow process but large enough to have been noticed since ancient times.

The measure of time has been based on the value of the angular rotation ω; as we have seen, its variation is governed by Eq. (19).

Another rewarding insight can be gained by considering the fundamental equation of the rotational motion; this specifies that the rate of change of the angular momentum vector $\mathbf{H}$ about the centroid shall be equal to the torque $\mathbf{G}$ exerted by the disturbing forces:

$$\frac{d\mathbf{H}}{dt} = \mathbf{G}. \tag{26}$$

This torque exists because of the uneven distribution of mass about the centroid due to the fact that the axis of rotation does not coincide with the axis of figure; this condition also causes the two moments of inertia A and C to differ. No other considerations of internal density distribution are made here; rigidity is tantamount to the fact that only the moments of inertia are of dynamical significance. The components of **H** along the principal axes of inertia (ξ, η, ζ) are $A\omega_1$, $A\omega_2$, $C\omega_3$. Since

$$\begin{aligned} \omega_1 &= \omega \sin\theta_r \cos\phi_r, \\ \omega_2 &= \omega \sin\theta_r \sin\phi_r, \\ \omega_3 &= \omega \cos\theta_r, \end{aligned} \tag{27}$$

we easily get

$$H^2 = \omega^2(A^2 \sin^2\theta_r + C^2\cos^2\theta_r) = H_0^2 + \omega^2 C^2 \cos^2\theta_r,$$

where H_0 is the projection of **H** onto the $(\xi\eta)$-plane. Also, from

$$H_0 = H\sin(\widehat{H\zeta}) = A(\omega_1^2 + \omega_2^2)^{1/2} = A\omega\sin\theta_r,$$

it follows that

$$\sin(\widehat{H\zeta}) = (A/H)\omega\sin\theta_r. \tag{28}$$

Denoting by v the angle between **H** and $\boldsymbol{\omega}$, we have

$$\begin{aligned} H\cos v &= (A\omega_1)\sin\theta_r\cos\phi_r + (A\omega_2)\sin\theta_r\sin\phi_r + (C\omega_3)\cos\theta_r \\ &= \omega(A\sin^2\theta_r + C\cos^2\theta_r), \end{aligned}$$

and this leads to

$$\sin v = [(C - A)/H]\omega\sin\theta_r\cos\theta_r. \tag{29}$$

We also have

$$\cos(\widehat{\xi H_0})\sin(\widehat{H\zeta}) = \cos(\widehat{\xi H}),$$

whence

$$\cos(\widehat{\xi H_0}) = \frac{A\omega_1}{H\sin(H\zeta)} = \frac{A\omega_1}{A\omega\sin\theta_r} = \cos\phi_r. \tag{30}$$

Similarly,

$$\cos(\widehat{\eta H_0}) = \frac{A\omega_2}{H\sin(\widehat{H\zeta})} = \frac{A\omega_2}{A\omega\sin\theta_r} = \sin\phi_r, \tag{31}$$

where Eqs. (27) and (28) have been used.

From Eqs. (30) and (31) it follows that the vectors **H** and $\boldsymbol{\omega}$ issued from the centroid O are coplanar with the axis of figure (ζ-axis) and that both lie on the same side of the ζ-axis. Furthermore, since $\sin(\widehat{H\zeta}) < \sin\theta_r$, **H** lies between the ζ-axis and $\boldsymbol{\omega}$.

The motion of **H** in inertial space depends entirely upon the external forces; only in the absence of such perturbing forces will the vector **H** remain fixed in inertial space. In either case, it is evident that the rotational motion of a rigid Earth will take place in such a manner that both the axis of figure and the rotational velocity vector will remain in the same plane with **H**, lying always on opposite sides of it.

Equation (23) can be interpreted as a rotation through the angle $\phi_r = \mu t - \sigma$ that this plane of the three vectors undergoes about the axis of figure ζ; since σ is not constant, this rotation is not uniform. During this motion, the vectors $\boldsymbol{\omega}$ and **H** describe conical surfaces within the Earth about the axis of figure which have the variable semiapertures θ_r and $\theta_r - v$, respectively.

Poinsot has given an elegant interpretation of the unperturbed rotational motion of a rigid body by showing that it can be generated by the rolling, without slipping, of a cone fixed within the rigid body (which is called the polhode cone) onto a cone which is fixed in inertial space, (which is called the herpolhode), where the line of contact between these two cones is the instantaneous axis of rotation for the rigid body.

In the case of Earth, the polhode cone has its vertex at O, its axis coincides with the axis of figure ζ, and $\theta_{r,0}$ is its semiaperture; the herpolhode cone also has O as its vertex, **H** as its axis, and v as its semiaperture. Since $\theta_{r,0}$ is a small angle, we have from Eq. (29)

$$v/\theta_{r,0} \sim (C - A)/C, \tag{32}$$

which shows that $v \ll \theta_{r,0}$: the polhode cone therefore has a much larger aperture than the herpolhode.

During a complete revolution in space, the axis of rotation $\boldsymbol{\omega}$ progresses along the larger cone fixed to the Earth through a distance equal to the circumference of the smaller space cone. Consequently, as $\boldsymbol{\omega}$ returns to the same position on the space cone, it will lie in a different position within the Earth, and the Earth will occupy a different position in space. The three aforementioned axes remain coplanar all the time, lying in a plane that rotates about **H**.

If the lunisolar forces are taken into consideration, then **H** moves in inertial space and this motion will drag along the vector $\boldsymbol{\omega}_0$ and the Earth with it. The lunisolar forces depend on the mean longitudes of the Moon and Sun and produce an increment $\theta_{r,1}$ to the angle that the rotational vector $\boldsymbol{\omega}_0$ makes with the axis of figure; the numerical magnitude of $\theta_{r,1}$

is $\sim 0''.01$. The new rotational vector $\boldsymbol{\omega}$ will rotate about the $\boldsymbol{\omega}_0$ vector according to a cone having $\theta_{r,1}$ as its semiaperture; at the same time $\boldsymbol{\omega}$ must also be on a conical surface having the z-axis of the inertial reference as its axis and the angle ϵ_1 as its semiaperture. We have then a small cone $(\boldsymbol{\omega}_0, \theta_{r1})$ rolling over a bigger cone $(\mathbf{z}, \epsilon_1)$, the instantaneous axis $\boldsymbol{\omega}$ being the generator of contact. The three axes $\mathbf{z}, \boldsymbol{\omega}_0$, and $\boldsymbol{\omega}$ need not remain coplanar during this motion.

The observed value for $\theta_r = \theta_{r,0} + \theta_{r,1}$ is of the order of $0''.5$, most of which, it is believed, is of geophysical origin, although a small lunisolar component cannot be discounted.

To a good approximation, we have

$$\nu/\theta_r = (C - A)/C, \qquad \nu/(\theta_r - \nu) = (C - A)/A.$$

Thus $\boldsymbol{\omega}$ virtually coincides with $\mathbf{H}$ and ω_3 constitutes nearly all of ω. The period of rotation of $\boldsymbol{\omega}$ about $\mathbf{H}$ is approximately one day, but the angular amplitude is so small that it leaves the celestial pole almost stationary among the stars. The variation of the axis of rotation with respect to the axis of figure causes a variation in astronomical latitude and longitude which is of importance in positional astronomy as well as in geodesy.

The motion of the celestial pole among the stars is due primarily to the lunisolar perturbation on $\mathbf{H}$; however, its variation $\theta_{r,1}$ of $\boldsymbol{\omega}$ from $\boldsymbol{\omega}_0$ is so small that there is no appreciable variation of the geographical pole of rotation during its diurnal circuit.

From Eqs. (27) and (3) we can define the location of the instantaneous rotational axis $\boldsymbol{\omega}$ with respect to the body axes as

$$\begin{aligned} \sin\theta_r \cos\phi_r &= (1/\omega)(\dot{\gamma}\cos\beta + \dot{\alpha}\sin\gamma\sin\beta), \\ \sin\theta_r \sin\phi_r &= (1/\omega)(-\dot{\gamma}\sin\beta + \dot{\alpha}\sin\gamma\cos\beta). \end{aligned} \tag{33}$$

$\dot{\alpha}$, $\dot{\gamma}$ are here the complete solutions of Eq. (9) consisting of the Eulerian and lunisolar components of the motion. As a first step we can approximate $\sin\theta_r$ by θ_r; we next remark that the Eulerian motion of α, γ is obtainable from Eq. (9) by setting $U = 0$ in it, and one can verify that it is given by Eq. (25) with the h and σ variables replaced by the arbitrary constants h_0, σ_0. We shall represent this Eulerian motion by means of

$$\theta_{r,0}\cos\phi_{r,0}, \qquad \theta_{r,0}\sin\phi_{r,0}. \tag{34}$$

The lunisolar component of the motion of α, γ can be approximated by Eq. (12), which we rewrite as

$$\begin{aligned} (\sin\gamma)\,\delta\alpha &= \theta_{r,1}\cos(\beta + \phi_{r,1}), \\ -\delta\gamma &= \theta_{r,1}\sin(\beta + \phi_{r,1}) \end{aligned}$$

in terms of the small angles $\delta\alpha = \alpha - \alpha_r$ and $\delta\gamma = \gamma - \epsilon_1$. Solving the above equation, we find

$$\begin{aligned} \theta_{r,1}\cos\phi_{r,1} &= (\sin\gamma)(\delta\alpha)\cos\beta - (\delta\gamma)\sin\beta, \\ -\theta_{r,1}\sin\phi_{r,1} &= (\sin\gamma)(\delta\alpha)\sin\beta + (\delta\gamma)\cos\beta, \end{aligned} \tag{35}$$

which can be effectively evaluated once we know the expressions for $\delta\alpha$ and $\delta\gamma$; these are provided by Eqs. (64) and (65) and were obtained as a by-product of the Poisson equation. Thus Eqs. (33) can be represented as the sum of two terms: the Eulerian component [Eq. (34)] and the lunisolar component [Eq. (35)].

The composite motion can be visualized as occurring along an epicycle. $\theta_{r,0}$, the radius of the main circle, is of the order 0".5 and represents the Eulerian motion of the pole of rotation with respect to the pole of figure with a period of $(2\pi/\omega_3)A/(C - A) = 305$ days, moving counterclockwise as seen from the north. The lunisolar component is a circle of radius $\theta_{r,1} \sim 0".01$ whose circumference moves clockwise according to a diurnal period about a center that in turn moves along the circumference of the Eulerian circle.

5.04 DEVELOPMENT OF THE DISTURBING FUNCTION. MOTION OF THE ECLIPTIC AND EQUINOX OF DATE

The force function U that describes the gravitational attraction of an external, rigid body of mass m' having equal principal moments of inertia is given by MacCullagh's formula:

$$U = G\left[\frac{m'm_1}{d} + \frac{m'}{2d^3}(A + B + C - 3I_d)\right] + \cdots. \tag{36}$$

Here m_1 is the Earth's mass; A, B, C are the Earth's principal moments of inertia; d is the distance of the center of mass O' of the disturbing body from the Earth's center of mass O; and I_d is the moment of inertia of the Earth about the line joining the centers.

The distance d between the two centers does not depend on the orientation of the Earth's principal axes of inertia $(O; \xi, \eta, \zeta)$ and will therefore be independent of the angular variables α, β, γ with respect to which one has to differentiate. As a consequence, we can neglect the term $m_1 m'/d$ within the expression for U. Denoting by ξ', η', ζ' the coordinates of O' with respect to the Earth's principal axes of inertia, we can write

$$I_{\mathrm{d}} = [A(\xi'^2 + \eta'^2) + C\zeta'^2]/d^2 = A + (C - A)\zeta'^2/d^2.$$

Replacing this expression within Eq. (36), we obtain

$$U = -\tfrac{3}{2}Gm' \frac{C - A}{d^5} (\zeta')^2, \tag{37}$$

as the only term to be retained for differentiation purposes. To obtain ζ' we must (1) invert the coordinate transformation given by Eq. (1) and write

$$\begin{bmatrix} \xi' \\ \eta' \\ \zeta' \end{bmatrix} = A^{\mathrm{T}} \begin{bmatrix} x' \\ y' \\ z' \end{bmatrix}, \tag{38}$$

where A^{T} is the transpose of the matrix A given by Eq. (2) (we note here in passing that since A is an orthogonal matrix, one must have $A^{-1} = A^{\mathrm{T}}$); and (2) represent the inertial coordinates (x', y', z') of O' in terms of β_0 and λ_0, which denote the latitude and longitude of O' referred to the fixed mean equinox and ecliptic of epoch. We get

$$\zeta' = d[\sin\gamma\cos\beta_0\sin(\alpha - \lambda_0) + \cos\gamma\sin\beta_0].$$

The Poisson equations [Eq. (10)] accordingly become

$$\begin{aligned} \dot{\gamma} &= \frac{3Gm'(C - A)}{C\omega_3}\frac{1}{d^3}[\sin\gamma\cos\beta_0\sin(\alpha - \lambda_0) \\ &\quad + \cos\gamma\sin\beta_0]\cos\beta_0\cos(\alpha - \lambda_0), \\ \dot{\alpha} &= -\frac{3Gm'(C - A)}{C\omega_3}\frac{1}{d^3}[\sin\gamma\cos\beta_0\sin(\alpha - \lambda_0) \\ &\quad + \cos\gamma\sin\beta_0][\cot\gamma\cos\beta_0\sin(\alpha - \lambda_0) - \sin\beta_0]. \end{aligned} \tag{39}$$

The solar and lunar theories provide the latitude and longitude (β^*, λ^*) for the center of mass of the Sun and the Moon referred to the mean equinox and ecliptic of date. One must therefore relate the two coordinate sets (β_0, λ_0) and (β^*, λ^*) by means of the inclination angle π_1 of the ecliptic of date onto the ecliptic of epoch and the longitude Π_1 of its ascending node on the ecliptic of epoch reckoned from the mean equinox of epoch. Figure 5.3 shows schematically the relative positions of the two fundamental planes. Λ is the longitude of the ecliptic of date ascending node (l_2) onto the ecliptic of epoch reckoned from the mean equinox of date (l_3).

A rotation of the inertial system $(O; x, y, z)$ about z-axis through the angle Π_1 followed by a rotation through the angle π_1 about the l_2-axis relate

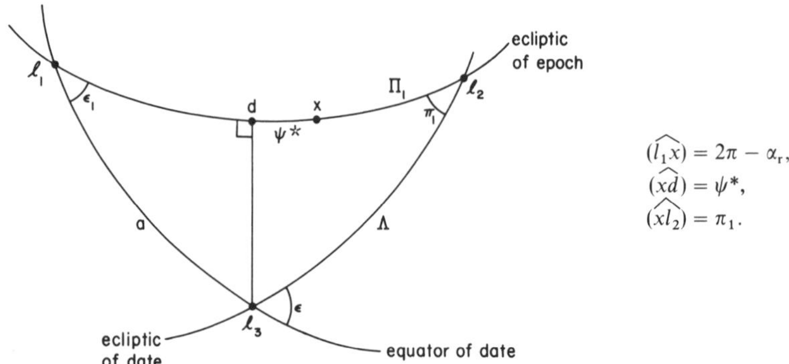

Fig. 5.3 Relative position of the ecliptic of date, equator of date, and ecliptic of epoch.

the three sets of angles (β_0, λ_0), (β^*, λ^*), and (π_1, Π_1) as follows:

$$\begin{bmatrix} \cos\beta_0 \cos(\lambda_0 - \Pi_1) \\ \cos\beta_0 \sin(\lambda_0 - \Pi_1) \\ \sin\beta_0 \end{bmatrix} = \begin{bmatrix} 1 & 0 & 0 \\ 0 & \cos\pi_1 & -\sin\pi_1 \\ 0 & \sin\pi_1 & \cos\pi_1 \end{bmatrix} \begin{bmatrix} \cos\beta^* \cos(\lambda^* - \Lambda) \\ \cos\beta^* \sin(\lambda^* - \Lambda) \\ \sin\beta^* \end{bmatrix}. \quad (40)$$

This matrix equation must be used to represent the expressions appearing on the right-hand sides of the Poisson equations in terms of β^* and λ^*, after having written the angle $\alpha - \lambda_0$ as the difference $(\alpha - \Pi_1) - (\lambda_0 - \Pi_1)$.

Before proceeding any further with the study of the Poisson equation, we plan to solve the spherical triangle $(l_1 l_2 l_3)$ of Fig. 5.3 for the two unknown sides Λ and $a = (l_1 l_3)$ and the unknown angle ϵ. Λ is required in Eq. (40); the other quantities are of interest in determining the motion of the equinox and obliquity of date with respect to be fixed reference frame. Using the Napier formulas, we get

$$\tan\left(\frac{\Lambda + a}{2}\right) = \frac{\cos\frac{1}{2}(\epsilon_1 - \pi_1)}{\cos\frac{1}{2}(\epsilon_1 + \pi_1)} \tan\left(\frac{\Pi_1 - \alpha_r}{2}\right),$$
$$\tan\left(\frac{\Lambda - a}{2}\right) = \frac{\sin\frac{1}{2}(\epsilon_1 - \pi_1)}{\sin\frac{1}{2}(\epsilon_1 + \pi_1)} \tan\left(\frac{\Pi_1 - \alpha_r}{2}\right). \quad (41)$$

Equations (41) are of the form

$$\tan x = h \tan y \quad (42)$$

with $x = (\Lambda \pm a)/2$ and $y = (\Pi_1 - \alpha_r)/2$; the parameter h is close to unity because π_1 is a small angle. As a consequence of Eq. (42), we can write

$$\tan(x - y) = \frac{\tan x - \tan y}{1 + \tan x \tan y} = \frac{(h - 1)\tan y}{1 + h \tan^2 y},$$

and consider the Fourier series expansion of

$$x - y = \arctan\left[\frac{(h-1)\tan y}{1 + h\tan^2 y}\right].$$

We can avoid evaluating the coefficients of the Fourier series by integration if we reduce the expression within brackets by trigonometric transformations to the more amenable form

$$\frac{n\sin(2y)}{1 - n\cos(2y)},$$

where $n = (h-1)/(h+1)$, and use a result tabulated in Jolley (1961). This gives the Fourier expansion

$$x = y + \frac{h-1}{h+1}\sin(2y) + \frac{1}{2}\left(\frac{h-1}{h+1}\right)^2 \sin(4y) + \cdots.$$

By adding the two expansions corresponding to $2x = \Lambda + a$ and $2x = \Lambda - a$, we get

$$\Lambda = \Pi_1 - \alpha_r + \delta\Lambda, \tag{43}$$

where

$$\begin{aligned}\delta\Lambda = {} & (\tan(\epsilon_1/2) - \cot(\epsilon_1/2))\tan(\pi_1/2)\sin(\Pi_1 - \alpha_r) \\ & + \tfrac{1}{2}(\tan^2(\epsilon_1/2) + \cot^2(\epsilon_1/2))\tan^2(\pi_1/2)\sin 2(\Pi_1 - \alpha_r) + \cdots.\end{aligned} \tag{44}$$

This expression for $\delta\Lambda$ can be developed in power series of α_r, the time t, and the angle $\Delta\epsilon = \epsilon_1 - \epsilon_0$, where ϵ_0 is the angle between the ecliptic and the equator of epoch.

Let us consider the various steps required to accomplish these various operations:

1. The sine and cosine of α_r must be expanded in powers of α_r.
2. We write

$$\cot(\epsilon_1/2) - \tan(\epsilon_1/2) = 2\cot\epsilon_1 = 2\cot(\epsilon_0 + \Delta\epsilon),$$
$$\cot^2(\epsilon_1/2) + \tan^2(\epsilon_1/2) = 2 + 4\cot^2(\epsilon_0 + \Delta\epsilon),$$

and expand these functions into power series of $\Delta\epsilon$.

3. $\tan(\pi_1/2)$ shall be represented in terms of $\tan\pi_1$ according to the formula

$$\tan(\alpha/2) = [(1 + \tan^2\alpha)^{1/2} - 1]/\tan\alpha = \tfrac{1}{2}\tan\alpha - \tfrac{1}{8}\tan^3\alpha + \cdots.$$

4. Terms of the form

$$(\tan \pi_1)^r \tfrac{\sin}{\cos}(s\Pi_1)$$

can be expressed in terms of time via the formulas of planetary theory

$$\begin{aligned} \tan \pi_1 \sin \Pi_1 &= p_1 T + p_2 T^2 + \cdots, \\ \tan \pi_1 \cos \Pi_1 &= q_1 T + q_2 T^2 + \cdots. \end{aligned} \tag{45}$$

To the second powers of the small parameters, we get

$$\begin{aligned} \Lambda = \Pi_1 - \alpha_r - (p_1 \cot \epsilon_0)T - \left(p_2 \cot \epsilon_0 - p_1 q_1 \frac{1 + \cos^2 \epsilon_0}{2 \sin^2 \epsilon_0}\right) T^2 \\ + (q_1 \cot \epsilon_0) T \alpha_r + \frac{p_1}{\sin^2 \epsilon_0} T(\Delta\epsilon) + \cdots. \end{aligned} \tag{46}$$

From Newcomb (1960) we can write

$$\begin{aligned} \pi_1 \sin \Pi_1 &= 4''\!.964T + 0''\!.1939T^2 + \cdots, \\ \pi_1 \cos \Pi_1 &= -46''\!.845T + 0''\!.0545T^2 + \cdots \end{aligned} \tag{47}$$

in Julian centuries from 1900. The quantities π_1 and $\tan \pi_1$ are numerically interchangeable within an accuracy of $5 \cdot 10^{-5}$ sec of arc. The mean obliquity at 1900 is $\epsilon_0 = 23^\circ\, 27'\, 8''\!.26$. With these numerical values for the parameters, Eq. (46) can be written as

$$\Lambda = \Pi_1 - \alpha_r - 11''\!.442T - 0''\!.453T^2 - 107''\!.981 T\alpha_r + 31''\!.339T(\Delta\epsilon) + \cdots. \tag{48}$$

Repeating a similar procedure, we can solve Eq. (41) for the other side, $a = (l_1 l_3)$, of the spherical triangle and get

$$\begin{aligned} a = (p_1 \operatorname{cosec} \epsilon_0)T + (p_2 \operatorname{cosec} \epsilon_0 - p_1 q_1 \cos \epsilon_0 \operatorname{cosec}^2 \epsilon_0)T^2 \\ - (q_1 \operatorname{cosec} \epsilon_0)T\alpha_r - (p_1 \cos \epsilon_0 \operatorname{cosec}^2 \epsilon_0)T(\Delta\epsilon) + \cdots. \end{aligned} \tag{49}$$

Using the previous numerical data, we can write

$$a = 12''\!.472T + 0''\!.493T^2 + 117''\!.705T\alpha_r - 28''\!.750T(\Delta\epsilon) + \cdots. \tag{50}$$

The true obliquity of date ϵ shall be obtained by writing

$$\begin{aligned} \sin \epsilon &= \sin \epsilon_1 \frac{\sin(\Pi_1 - \alpha_r)}{\sin \Lambda} \\ &= \sin \epsilon_1 \left[1 + \frac{\sin(\Pi_1 - \alpha_r) - \sin \Lambda}{\sin \Lambda}\right]. \end{aligned}$$

If we make use of Eq. (43) and expand the expression within brackets in powers of α_r and $\delta\Lambda$, then we obtain

$$\begin{aligned}\sin\epsilon = (\sin\epsilon_1)\{1 - (\delta\Lambda)\cot\Pi_1 + \tfrac{1}{2}(\delta\Lambda)^2 - \alpha_r(\delta\Lambda) \\ + (\cot^2\Pi_1)[(\delta\Lambda)^2 - \alpha_r(\delta\Lambda)]\} + \cdots .\end{aligned} \tag{51}$$

From Eq. (45) we obtain

$$\cot\Pi_1 = \frac{q_1}{p_1}\left[\frac{1 + (q_2/q_1)T + \cdots}{1 + (p_2/p_1)T + \cdots}\right] = \frac{q_1}{p_1}\left[1 + \left(\frac{q_2}{q_1} - \frac{p_2}{p_1}\right)T + \cdots\right].$$

Using the above expression for $\cot\Pi_1$ and Eq. (46) for $\delta\Lambda$ and substituting into Eq. (51), we have

$$\sin\epsilon = \sin\epsilon_1(1 + S),$$

where

$$\begin{aligned}S = (q_1\cot\epsilon_0)T + (q_2\cot\epsilon_0 + \tfrac{1}{2}p_1^2\cot^2\epsilon_0 - \tfrac{1}{2}q_1^2)T^2 \\ + (p_1\cot\epsilon_0)T\alpha_r - (q_1\operatorname{cosec}^2\epsilon_0)T(\Delta\epsilon) + \cdots .\end{aligned} \tag{52}$$

We can then write

$$S\sin\epsilon_1 = \sin\epsilon - \sin\epsilon_1 = 2\sin\left(\frac{\epsilon - \epsilon_1}{2}\right)\cos[\epsilon_1 + \tfrac{1}{2}(\epsilon - \epsilon_1)],$$

and upon expanding these two functions around zero and ϵ_1 with increments equal to $\frac{1}{2}(\epsilon - \epsilon_1)$, we arrive at the expression

$$S\sin\epsilon_1 = (\cos\epsilon_1)[(\epsilon - \epsilon_1) - \tfrac{1}{2}(\epsilon - \epsilon_1)^2\tan\epsilon_1 + \cdots],$$

whence

$$\epsilon - \epsilon_1 = \tfrac{1}{2}(\epsilon - \epsilon_1)^2\tan\epsilon_1 + S\tan\epsilon_1 + \cdots,$$

and upon one iteration we get

$$\epsilon - \epsilon_1 = S\tan\epsilon_1 + \tfrac{1}{2}S^2\tan^3\epsilon_1 + \cdots . \tag{53}$$

Since $\epsilon_1 = \epsilon_0 + \Delta\epsilon$ and

$$\tan\epsilon_1 = \tan\epsilon_0 + \sec^2\epsilon_0(\Delta\epsilon) + \cdots,$$

we get from Eqs. (53) and (52)

$$\epsilon = \epsilon_0 + \Delta\epsilon + q_1 T + (\tfrac{1}{2}p_1^2\cot\epsilon_0 + q_2)T^2 + p_1 T\alpha_r + \cdots .$$

Using the numerical values for the constant parameters, we have

$$\epsilon = \epsilon_0 + \Delta\epsilon - 46''.845T + 0''.05463T^2 + 4''.964T\alpha_r + \cdots . \tag{54}$$

Another quantity of interest is the longitude of the mean equinox of date referred to the fixed mean equinox and ecliptic of epoch; this is the arc$(\widehat{xd})$ in Fig. 5.3, denoted by ψ^* and measured westward from the equinox of epoch. From the right spherical triangle $(l_1 d l_3)$ of Fig. 5.3, we have

$$\begin{aligned}\tan(\widehat{l_1 d}) &= \tan(2\pi - \alpha_r - \psi^*) = -\tan(\alpha_r + \psi^*) \\ &= (\tan a)(\cos\epsilon_1).\end{aligned} \tag{55}$$

From

$$\cos\epsilon = \cos\pi_1 \cos\epsilon_1 - \sin\pi_1 \sin\epsilon_1 \cos(\Pi_1 - \alpha_r),$$

using the small angle approximation for π_1 and $\epsilon - \epsilon_1$, we can write

$$\epsilon - \epsilon_1 = \pi_1 \cos(\Pi_1 - \alpha_r). \tag{56}$$

Next, from

$$\sin a = \frac{\sin\pi_1}{\sin\epsilon} \sin(\Pi_1 - \alpha_r),$$

we approximate to get

$$\begin{aligned}\tan a \sim \sin a &= \frac{\pi_1 \sin(\Pi_1 - \alpha_r)}{\sin\epsilon_1} [1 + (\epsilon - \epsilon_1)\cot\epsilon_1]^{-1} \\ &= \frac{\pi_1}{\sin\epsilon_1} \sin(\Pi_1 - \alpha_r)[1 - \pi_1 \cot\epsilon_1 \cos(\Pi_1 - \alpha_r)],\end{aligned} \tag{57}$$

where use has been made of Eq. (56). From Eqs. (55) and (57), we have

$$\begin{aligned}-\tan(\alpha_r + \psi^*) &= \tan a \cos\epsilon_1 \\ &= \pi_1 \sin(\Pi_1 - \alpha_r)\cot\epsilon_1 \\ &\quad - \pi_1^2 \sin(\Pi_1 - \alpha_r)\cos(\Pi_1 - \alpha_r)\cot^2\epsilon_1.\end{aligned}$$

Comparing the above expression with Eqs. (43) and (44), we have

$$\Lambda = \Pi_1 - \alpha_r + \tan(\alpha_r + \psi^*) + \tfrac{1}{2}\pi_1^2 \sin(\Pi_1 - \alpha_r)\cos(\Pi_1 - \alpha_r).$$

As an approximation we can write

$$\Lambda = \Pi_1 + \psi^* + \tfrac{1}{2}\pi_1^2 \sin\Pi_1 \cos\Pi_1.$$

Using Eq. (45), we reach the final result:

$$\psi^* = \Lambda - \Pi_1 - \tfrac{1}{2} p_1 q_1 T^2. \tag{58}$$

Formulas (48), (50), (54), and (58) provide an approximate representation of those elements that describe the motion of the ecliptic and equinox of date with reference to the inertial frame. These reduction formulas between the two frames are essential in order to update the precessional and nutational motion of Earth given at a fixed epoch.

5.05 SOLUTION OF THE POISSON EQUATIONS IN TERMS OF THE LUNAR AND SOLAR THEORIES

We proceed with the task of representing the right-hand sides of the two Poisson equations (39) in terms of the small parameters that pertain to perturbation theory.

The angle γ will be written as $\gamma = \epsilon_0 + \Delta\epsilon + \gamma - \epsilon_1$, where $\Delta\epsilon = \epsilon_1 - \epsilon_0$; the functions $\sin\gamma$, $\cos\gamma$, $\cot\gamma$ will be developed in double power series of the small parameters $\Delta\epsilon$ and $\gamma - \epsilon_1$.

The transformation of coordinates represented by Eq. (40) will be used to express the coordinates β_0, λ_0 in terms of β^*, λ^* (for which tabular values are known) and of the angle Λ, which, according to Eq. (48), can be expanded in terms of T, $\Delta\epsilon$, α_r.

The terms containing $\cos\pi_1$ and $\sin\pi_1$ must be transformed to terms in $\tan\pi_1$ according to the elementary representations

$$\cos\alpha = (1 + \tan^2\alpha)^{-1/2} = 1 - \tfrac{1}{2}\tan^2\alpha + \tfrac{3}{8}\tan^4\alpha - \cdots,$$
$$\sin\alpha = (\tan\alpha)(1 + \tan^2\alpha)^{-1/2} = \tan\alpha - \tfrac{1}{2}\tan^3\alpha + \cdots.$$

This reduction will introduce secular terms in T through Eq. (45) of planetary theory.

These considerations prove in effect that it is possible to express the right-hand sides of the Poisson equations in terms of the latitude β^* and longitude λ^* of the disturbing body (Moon and Sun) referred to the true equinox and ecliptic of date. The coefficients of these coordinates are power series in T, $\Delta\epsilon$, α_r depending on the parameters p_i, q_i, ϵ_0.

From Newcomb's results on planetary perturbations we have

$$\begin{aligned} \alpha_r &= -5037''.19T + 1''.0721T^2, \\ \Delta\epsilon &= 9''.210\cos\Omega + 0''.551\cos 2L' + 0''.0606T^2, \\ \alpha_r^2 &= 123''.0132T^2, \end{aligned} \tag{59}$$

where Ω is the mean longitude of the Moon's ascending node and L' is the Sun's mean longitude. Replacing these approximations in the previous power

Table 5.1

Right-Hand Sides of Poisson Equations Exclusive of Common Multiplier: Lunar and Solar Contributions

l	l'	F	D	Ω	$\dot{\alpha}$-eq. cosine terms (rad · 10^6)	$\dot{\gamma}$-eq. sine terms (rad · 10^6)
					Lunar Contributions	
0	0	0	0	0	455265	
0	0	0	2	0	12354	
0	0	2	0	0	5473	
1	0	0	−2	0	14243	
1	0	0	0	0	74480	
1	0	0	2	0	1960	
1	0	2	0	0	1050	
2	0	0	0	0	6107	
−1	0	0	0	1	6295	−3365
−1	0	0	2	1	1334	
−1	0	2	0	1	2113	−1130
−1	0	2	2	1	−2772	1482
0	0	0	−2	1	1031	
0	0	0	0	1	77029	−41166
0	0	0	2	1	−1148	
0	0	2	−2	1	2130	−1138
0	0	2	0	1	−76215	40736
0	0	2	2	1	−2330	1245
1	0	0	−2	1	1205	
1	0	0	0	1	6373	−3405
1	0	2	0	1	−14587	7797
2	0	2	0	1	−1932	1032
−1	0	2	0	2	12812	−5558
−1	0	2	2	2	−16479	7149
0	−1	2	0	2	−1376	
0	0	0	0	2	−1867	
0	0	2	0	2	−453156	196588
0	0	2	2	2	−13857	6012
0	1	2	0	2	1563	
1	0	2	−2	2	3343	−1450
1	0	2	0	2	−86762	37639
1	0	2	2	2	−3348	1453
2	0	2	−2	2	1362	
2	0	2	0	2	−11482	4981
3	0	2	0	2	−1294	
					Solar Contributions	
0	0	0	0	0	458887	
0	1	0	0	0	23058	
0	0	2	−2	2	−458374	198853
0	1	2	−2	2	−26876	11659
0	2	2	−2	2	−1094	
0	1	−2	2	−2	3842	1667

series developments, we can transform Eqs. (39) into

$$\dot{\gamma} = \frac{3Gm'(C - A)}{C\omega_3} \frac{1}{d^3} [(0.917392) \cos \beta^* \sin \beta^* \cos \lambda^*$$
$$+ (0.397985 - 0.0002t) \cos^2 \beta^* \sin \lambda^* \cos \lambda^*],$$
$$\dot{\alpha} = -\frac{3Gm'(C - A)}{C\omega_3} \frac{1}{d^3} [(1.716684 + 0.0008t \tag{60}$$
$$- 0.0003 \cos \Omega) \cos \beta^* \sin \beta^* \sin \lambda^* + (0.0001t) \cos \beta^* \sin \beta^* \cos \lambda^*$$
$$+ (0.917392 - 0.0003t)(\cos^2 \beta^* \sin^2 \lambda^* - \sin^2 \beta^*)].$$

Using Brown's lunar theory and Newcomb's results on solar perturbations, one can expand expressions like $(c/d)^3 \cos^2 \beta^* \sin \beta^* \sin \lambda^*$ which appear in the right-hand sides of Eqs. (60) in terms of the cosines and sines of the fundamental angular variables (l, l', F, D, Ω), see Chapter VI. Here c denotes Earth's mean distance from the disturbing body.

The detailed expansions undertaken by Woolard (1953) contain also terms in $T\cos(\ )$ and $T\sin(\ )$. Listed in Table 5.1 are only those terms whose amplitudes reach the value of 10^{-3} radians. The right-hand side of the $\dot{\alpha}$-equation contains only cosine terms; for the $\dot{\gamma}$-equation we have only sine terms. Both lunar and solar contributions are listed in Table 5.1.

We finally are in a position to integrate the Poisson equations by making use of the results listed in Table 5.1; one must know, however, the motions of the fundamental arguments (l, l', F, D, Ω), these we shall denote $n_1, \ldots, n_5$, respectively. Thus, e.g., a term of the form $A \sin(al + bF)$ listed in Table 5.1 will generate the term $A \sin[(an_1 + bn_3)t]$, and by integration one obtains the term $-[A/(an_1 + bn_3)] \cos(al + bF)$.

The motions of the fundamental arguments for the epoch 1900, expressed in radians per Julian century (J.c.), are

$$\begin{aligned} n_1 &= 8328.691233, \\ n_2 &= 628.301946, \\ n_3 &= 8433.466421, \\ n_4 &= 7771.377324, \\ n_5 &= -33.757146. \end{aligned} \tag{61}$$

No secular rates for these quantities need be used for the purpose of the above-mentioned integration.

To evaluate the common multiplier $3Gm'(C - A)/C\omega_3 c^3$, where m' is the mass of the Moon, we note that in the expansion for $\dot{\gamma}$ there is the term

$-41{,}166 \times 10^{-6} \sin \Omega$, which is due solely to the Moon, and that by integration it will give rise to

$$\gamma = -\frac{3Gm'(C - A)}{C\omega_3 c^3} \frac{0.041166}{33.757146} \cos \Omega + \cdots .$$

The coefficient of $\cos \Omega$ in the above expression must coincide with the constant of nutation adopted for the epoch ($=9''\!.210$), as it appears from Eqs. (59). Thus for the lunar perturbations we must have

$$\frac{3Gm'(C - A)}{C\omega_3 c^3} = -(9''\!.210)\frac{33.757146}{0.041166} = -7552''\!.4295, \tag{62}$$

expressed in seconds of arc per J.c.

To determine the corresponding multiplier for the solar perturbations, we proceed as follows. The two constant terms appearing in Table 5.1 for the expansion of the $\dot{\alpha}$-equation are 0.455265, for the lunar contribution, and 0.458887 for the solar contribution. These two terms give rise to a secular term $\alpha = -PT + \cdots$, whose coefficient P must coincide with the adopted value of $5037''\!.19$ per J.c. for the lunisolar precession in longitude, as shown by Eqs. (59). Using Eq. (62), we can subtract the lunar contribution to this precessional rate, which amounts to

$$(7552''\!.4295)(0.455265) = 3438''\!.36.$$

This leaves an amount of $1598''\!.83$ per J.c. as the solar contribution to the precessional rate, and this then is due to the coefficient 0.458887. Thus we must have

$$\frac{3Gm'(C - A)}{C\omega_3 c^3} = \frac{1598''\!.83}{0.458887} = 3484''\!.15 \qquad \text{per J.c.,} \tag{63}$$

where m' stands for the solar mass and c for the astronomical unit.

The results of the integration depend on (1) the coefficients given in Table 5.1, (2) the numerical values of the common multipliers, and (3) the motions of the fundamental arguments. These results are summarized in Table 5.2, which provides the lunisolar precession in longitude (α-equation) and nutation in obliquity (γ-equation) referred to the fixed ecliptic of epoch. The α-equation contains only sine terms, the γ-equation only cosine terms. The secular term for α is $5037''\!.19$ per J.c.; the periodic terms are furnished by Table 5.2. Woolard (1953) carried calculations out to the order of 10^{-6} and found that $\dot{\gamma}$ contains the constant term 3.39×10^{-6} which is due primarily to the term $1''\!.53 \cos(F + \Omega)$ of the Moon's latitude β^*. By integration one reaches the secular term $-0''\!.0256T$ in obliquity. This term is primarily due to the secular motion of the ecliptic and the planetary disturbance in the Moon's motion.

Table 5.2

Lunisolar Precession and Nutation in Longitude and Obliquity Referred to Fixed Ecliptic (Julian Date 2415020.0).

l	l'	F	D	Ω	Longitude—sine terms	Obliquity—cosine terms
0	0	0	0	1	$-17''.2335$	$9''.2100$
0	0	2	−2	2	$-1''.2729$	$0''.5522$
0	0	0	0	2	$0''.2088$	$-0''.0904$
0	0	2	0	2	$-0''.2037$	$0''.0883$
0	1	0	0	0	$0''.1261$	
1	0	0	0	0	$0''.0675$	
0	1	2	−2	2	$-0''.0497$	$0''.0216$
0	0	2	0	1	$-0''.03419$	$0''.0183$
1	0	2	0	2	$-0''.02608$	$0''.0113$
0	−1	2	−2	2	$0''.0214$	
1	0	0	−2	0	$-0''.0149$	
−1	0	2	0	2	$0''.0114$	
0	0	2	−2	1	$0''.0124$	

Expressions for precession and nutation when referred to the ecliptic of date can be obtained by applying a correction to the corresponding quantities referred to the ecliptic of epoch. Equations (58) and (48) must be used to obtain the correction for the precession in longitude, whereas Eq. (54) will be used to correct the nutation of the obliquity. The above-listed formulas provide corrections up to the second powers of T, α_r, $\Delta\epsilon$; the values for these last two parameters must be replaced from Table 5.2. In a first-order approximation, we have a correction of $-11''.442T$ for the precession in longitude and a correction of $-46''.845T$ for the nutation of the obliquity.

Equations (60) are equivalent to Eqs. (10); however, one should emphasize the fact that in writing the right-hand sides of Eqs. (10) a certain number of approximations had to be made before one could reach Eqs. (60); specifically: (1) higher-order terms in Earth's gravitational potential were ignored; (2) the disturbing body was assumed to be a homogeneous sphere; (3) the precessional and nutational expansions from Newcomb's theory were also approximated; and (4) only the Moon's and the Sun's perturbing effects were taken into account.

The Poisson equations are good approximations not only to Eqs. (9), which describe the motion of the body axes with respect to the inertial frame, but also to Eqs. (17), which determine the motion of the instantaneous rotational axis with respect to the body axes.

In seeking solutions to Eqs. (9) one must consider the fact that these equations differ from the Poisson equations by terms which are multiplied by the factor $A/C\omega_3$, which numerically is approximately 4.32×10^{-6} J.c.

Similarly, we note that Eqs. (17) differ from (10) by terms which are multiplied by $(C - A)/C\omega_3$, which is 10^{-3} times smaller than the previous factor. It is legitimate therefore to represent $\dot{\alpha}$ and $\dot{\gamma}$ appearing in the right-hand sides of Eqs. (9) and (17) by the values provided in Table 5.1 multiplied by the appropriate common factors.

Both Eqs. (9) and (17) are more general than Eq. (10) because they do not vanish with U. Their general solutions must include the Eulerian motion besides the lunisolar motion. If we integrate equations similar to Eq. (25) but with constant h_0 and σ_0 to obtain the Eulerian motion, then we can solve Eq. (9) by the following formulas:

$$\begin{aligned}\delta\gamma &= -\frac{Ah_0}{C\omega_3}\sin\left(\frac{C\omega_3}{A}t - \sigma_0\right) + \frac{A}{C\omega_3}\left[\dot{\alpha}\sin\gamma + \int(\cos\gamma)(\dot{\alpha}\dot{\gamma})\,dt\right],\\ \delta\alpha &= \frac{Ah_0}{C\omega_3\sin\gamma}\cos\left(\frac{C\omega_3}{A}t - \sigma_0\right) + \frac{A}{C\omega_3}\left[\int(\cos\gamma)(\dot{\alpha})^2\,dt - \frac{\ddot{\gamma}}{\sin\gamma}\right],\end{aligned} \tag{64}$$

where $h_0 = \omega_0 \sin\theta_{\mathrm{r},0} \sim \omega_3\theta_{\mathrm{r},0}$ with $\theta_{\mathrm{r},0} \sim 0''.5$. Using the results in Table 5.1, we reach the following numerical formulation:

$$\begin{aligned}\delta\gamma = {} & -\frac{A}{C}\theta_{\mathrm{r},0}\sin\left(\frac{C\omega_3}{A}t - \sigma_0\right) - 0''.00868\\ & + 0''.00590\cos(2F + 2\Omega) + 0''.00113\cos(l + 2F + 2\Omega)\\ & - 0''.00100\cos\Omega + 0''.00275\cos(2F - 2D + 2\Omega) + \cdots,\\ \delta\alpha = {} & \frac{A}{C}\frac{\theta_{\mathrm{r},0}}{\sin\gamma}\cos\left(\frac{C\omega_3}{A}t - \sigma_0\right) + 0''.01615\sin(2F + 2\Omega)\\ & - 0''.00338\sin\Omega + 0''.00334\sin(2F + \Omega)\\ & + 0''.00309\sin(l + 2F + 2\Omega) + 0''.00753\sin(2F - 2D + 2\Omega) + \cdots.\end{aligned} \tag{65}$$

5.06 DYNAMICS OF A SYSTEM OF TWO RIGID BODIES OF COMPARABLE SIZE

In the previous section we have seen the application of Brown's lunar theory and Newcomb's solar theory for the purpose of determining the perturbations caused by both Moon and Sun upon the motion of Earth's axis. However, in doing so, we had to make some simplifying assumptions so as to arrive at a feasible solution. The most stringent of these assumptions concerned the constitution of the disturbing body.

In this section we want to develop an alternative approach in which we assume that the perturbing body is comparable in size and constitution to the main body. We shall briefly examine the dynamics of an isolated binary system of rigid bodies.

More specifically, we are assuming that each body of the system is a rigid, homogeneous oblate ellipsoid of revolution, having mass m_i, equatorial radius a_i, and polar radius $c_i < a_i$ for $i = 1, 2$. We denote by e_i the eccentricity of the elliptical meridian and assume that the square of this quantity

$$\epsilon_i = (e_i)^2 = 1 - (c_i/a_i)^2 \tag{66}$$

is a small parameter.

The body axes $(O_i; \xi_i, \eta_i, \zeta_i)$ of each ellipsoid have origin at the center of symmetry O_i; the ξ_i- and η_i- axes are perpendicular to each other in the equatorial plane, whereas the ζ_i-axis coincides with the polar axis.

The moment of inertia with respect to any equatorial axis passing through the centroid is given by

$$A_i = \tfrac{1}{5}m_i(a_i)^2(2 - \epsilon_i), \tag{67}$$

while the moment of inertia of each ellipsoid with respect to the polar axis is

$$C_i = \tfrac{2}{5}m_i(a_i)^2. \tag{68}$$

The ratio of these two moments is then

$$C_i/A_i = 2/(2 - \epsilon_i) = \sum_{k=0}^{\infty} (\epsilon_i/2)^k. \tag{69}$$

We introduce an inertial, orthogonal coordinate reference with origin at the centroid O of the two ellipsoids, with the x-, y-, and z-axes fixed in inertial space. The relative position of the two centroids with respect to the inertial frame can be expressed by means of cylindrical coordinates ρ, z, ϕ, the longitude ϕ being measured from the x-axis; see Fig. 5.4.

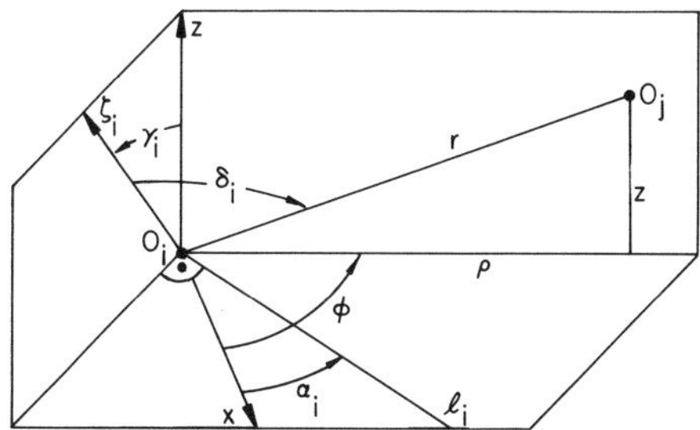

Fig. 5.4 Inertial frame which is used to study the linear motion of the two centroids and the rotational motion for each set of body axes for the components of a binary system. δ is the libration angle.

Denoting by r the variable distance (O_1O_2), we have $r^2 = \rho^2 + z^2$. The orientation of the body axes of each ellipsoid with respect to the inertial frame is defined by the three Euler angles α_i, β_i, γ_i, as in Fig. 5.1. Another angular variable of great interest is the libration angle δ_i between the polar axis ζ_i and the line O_iO_j of the two centroids. One gets

$$\nu_i = \cos\delta_i = \cos(\zeta_i O_i O_j) = (z/r)\cos\gamma_i + (\rho/r)\sin\gamma_i\sin(\alpha_i - \phi). \tag{70}$$

We denote by ω_{ik} $(i = 1, 2, k = 1, 2, 3)$ the three components of the rotational velocity vector $\boldsymbol{\omega}_i$; they satisfy relations analogous to Eq. (3).

The equations governing the motion of the binary system are obtainable by expressing the Lagrangian of the system,

$$L = T_0 + T_1 + T_2 + V^*,$$

in terms of the nine parameters of motion ρ, z, ϕ, α_i, β_i, γ_i, with $i = 1, 2$.

T_0 is the kinetic energy of the two point masses and is given by

$$2T_0 = \frac{m_1 m_2}{m_1 + m_2}(\dot{\rho}^2 + \rho^2\dot{\phi}^2 + \dot{z}^2).$$

T_1 and T_2 represent the kinetic energy of each body which is due solely to its own rotation; they are given by Eq. (8).

V^* is the total potential energy of the system and is obtainable by integrating the expressions for the internal and external potentials due to each mass over the volume occupied by both masses and averaging the contributions from the two components of the system.

The integral of the internal potential extended over the volume occupied by the generating mass itself is

$$4\pi\sigma CG\left(\frac{1-\epsilon}{\epsilon}\right)^{1/2} \arctan\left(\frac{\epsilon}{1-\epsilon}\right)^{1/2},$$

where σ is the constant density. This expression does not contain any of the parameters of motion and will not contribute to the final equations of motion.

The contribution which is obtained when integrating the exterior potential generated by each mass over the volume occupied by the other mass and then averaging over the whole system can be expressed as an infinite power series in the two small parameters ϵ_1, ϵ_2 with coefficients depending on the two libration angles δ_1, δ_2 and also on the angle between the two rotational axes whose cosine is given by

$$\mu = \cos(\zeta_1\zeta_2) = \cos\gamma_1\cos\gamma_2 + \sin\gamma_1\sin\gamma_2\cos(\alpha_2 - \alpha_1). \tag{71}$$

The leading terms of V^* are

$$V^* = \frac{Gm_1m_2}{r}\left\{1 - \sum_{k=1}^{2}\left[\frac{1}{5}\left(\frac{a_k}{r}\right)^2 P_2(v_k)\epsilon_k + \frac{3}{35}\left(\frac{a_k}{r}\right)^4 P_4(v_k)\epsilon_k^2\right]\right.$$
$$+ \frac{1}{25}\epsilon_1\epsilon_2\left(\frac{a_1}{r}\right)^2\left(\frac{a_2}{r}\right)^2\left[P_2(\mu) + 15\mu v_1 v_2\right.$$
$$\left.\left. + \frac{45}{4}(v_1^2 + v_2^2) - \frac{5}{2}\right]\right\} + \cdots, \tag{72}$$

where the P are Legendre polynomials.

The general formula for V^* was obtained by Lanzano (1969); it was expressed by means of special spherical harmonics consisting of the product of Gegenbauer polynomials and biaxial harmonics.

Using the previous expressions for the various components of the Lagrangian, we can write the equations of motion:

$$\begin{aligned}
\ddot{\rho} - \rho(\dot{\phi})^2 &= \frac{m_1 + m_2}{m_1 m_2}\frac{\partial V^*}{\partial \rho}, \\
\ddot{z} &= \frac{m_1 + m_2}{m_1 m_2}\frac{\partial V^*}{\partial z}, \\
\frac{d}{dt}(\rho^2\dot{\phi}) &= \frac{m_1 + m_2}{m_1 m_2}\frac{\partial V^*}{\partial \phi}, \\
\dot{\alpha}_i \cos\gamma_i + \dot{\beta}_i &= \omega_{i3} = \text{const.}, \\
\frac{d}{dt}\left(\dot{\alpha}_i \sin^2\gamma_i + \frac{C_i}{A_i}\omega_{i3}\cos\gamma_i\right) &= \frac{1}{A_i}\frac{\partial V^*}{\partial \alpha_i}, \\
\ddot{\gamma}_i - (\dot{\alpha}_i)^2 \sin\gamma_i\cos\gamma_i + \frac{C_i}{A_i}\omega_{i3}\dot{\alpha}_i\sin\gamma_i &= \frac{1}{A_i}\frac{\partial V^*}{\partial \gamma_i}.
\end{aligned} \tag{73}$$

The last three equations of the set are valid for $i = 1, 2$. The fourth equation of the set is already a first integral of motion. The above set constitutes a simultaneous nonlinear system of seven differential equations: three equations pertain to the relative motion of the two centroids; two equations describe the rotational motion of the body axes for each body of the system. We shall briefly study the above system and represent its solutions as power series of ϵ_1, ϵ_2.

In the zero-order approximation, that is to say, when we neglect the meridional eccentricities, the equations pertaining to the relative motion of the centroids (the so-called translational motion) are independent of those defining the motion of the body axes (i.e., the rotational motion).

The zero-order approximation for the rotational motion of each body satisfies the following equations:

$$\begin{aligned}\dot{\alpha}_0 \cos\gamma_0 + \dot{\beta}_0 = \omega_3 = \text{const.},\\ \dot{\alpha}_0 \sin^2\gamma_0 + \omega_3 \cos\gamma_0 = \text{const.},\\ (\dot{\gamma}_0)^2 + (\dot{\alpha}_0 \sin\gamma_0)^2 = \text{const.},\end{aligned} \tag{74}$$

where the subscript zero refers to the zero powers of ϵ_1, ϵ_2. We choose the particular solution

$$\dot{\alpha}_0 = \text{const.}, \qquad \dot{\beta}_0 = \text{const.}, \qquad \dot{\gamma}_0 = 0. \tag{75}$$

It consists of a constant-rate precession about the z-axis of the inertial frame that keeps the nutational angle unchanged and of a constant-rate rotation about the ζ-body axis. This solution agrees with the unperturbed Eulerian motion of Earth's axis.

The translational motion of the two centroids, in the zero-order approximation, reduces to the Kepler motion of two point masses. It is appropriate therefore to choose an elliptical motion of eccentricity e_0, semimajor axis A_0, and argument of periastron ω_0. With no loss of generality, we shall suppose that the plane of the Kepler orbit coincides with the (xy)-plane of the inertial frame. This gives rise to

$$\begin{aligned}z_0 = 0, \qquad r_0 = \rho_0, \quad \phi_0 = f + \omega_0,\\ \rho_0^2\dot{\phi}_0 = nA_0^2(1 - e_0^2)^{1/2} = \text{const.},\end{aligned} \tag{76}$$

where f is the true anomaly and n the mean motion of the Kepler orbit. In what follows we use the mean anomaly

$$M = n(t - \tau), \tag{77}$$

where τ is the time passage at periastron, as the new independent variable instead of the time t. We denote by primes the derivatives with respect to M.

By expanding the various terms appearing in Eq. (73) into power series of ϵ_1, ϵ_2, we obtain the various approximations for the governing equations of motion. The approximation of order (r, s) for the rotational motion is

$$\begin{aligned}(\alpha'_{rs} \sin^2\gamma_0 - E_0\gamma_{rs} \sin\gamma_0)' = A_l(M),\\ \gamma''_{rs} + \alpha'_0(\alpha'_0 + E_0 \cos\gamma_0)\gamma_{rs} + E_0\alpha'_{rs} \sin\gamma_0 = \Gamma_l(M),\end{aligned} \tag{78}$$

where $E_0 = \beta'_0 - \alpha'_0 \cos\gamma_0$ and the right-hand sides contain only lower-order terms.

These are linear differential equations with constant coefficients depending on the zero-order approximation. By integrating the first equation of the set between 0 and M we get

$$\alpha'_{rs}\sin\gamma_0 = E_0\gamma_{rs} + \frac{1}{\sin\gamma_0}\int_0^M A_l(\bar{M})\,d\bar{M}. \tag{79}$$

The choice of $M = 0$ as the initial value of the independent variable corresponds to the time $t = \tau$ of passage of the centroid through the periastron position.

Upon replacing this result within the second equation of the set (78), we have

$$\gamma''_{rs} + \Lambda_0^2\gamma_{rs} = \Gamma_l(M) - \frac{E_0}{\sin\gamma_0}\int_0^M A_l(\bar{M})\,d\bar{M} = F_l(M). \tag{80}$$

The quantity

$$\Lambda_0^2 = (\alpha'_0)^2 + E_0\beta'_0 \tag{81}$$

is always a positive number. Only the right-hand sides of these equations vary with the given approximation. The particular solution of Eq. (80) satisfying the initial conditions

$$\gamma(0) = \gamma'(0) = 0$$

can be easily written as

$$\Lambda\gamma = \int_0^M F(\bar{M})\sin[\Lambda(M - \bar{M})]\,d\bar{M}. \tag{82}$$

Following a similar procedure, one can write the approximation of order (r, s) for the relative motion of the two centroids as

$$\begin{aligned}(\rho_0^2\phi'_{rs} + 2\rho_0\phi'_0\rho_{rs})' &= \Phi_l(M),\\ \rho''_{rs} - [(\phi'_0)^2 + 2(A_0/\rho_0)^3]\rho_{rs} - 2\rho_0\phi'_0\phi'_{rs} &= P_l(M),\\ z''_{rs} + (A_0/\rho_0)^3 z_{rs} &= Z_l(M),\end{aligned} \tag{83}$$

where the right-hand sides depend on lower-order terms alone. Integration of the first equation yields

$$\phi'_{rs} = -2\frac{\phi'_0}{\rho_0}\rho_{rs} + \frac{1}{\rho_0^2}\int_0^M \Phi_l(\bar{M})\,d\bar{M}, \tag{84}$$

which when substituted in the second equation gives rise to

$$\rho''_{rs} + [3(\phi'_0)^2 - 2(A_0/\rho_0)^3]\rho_{rs} = 2(\phi'_0/\rho_0)\int_0^M \Phi_l(\bar{M})\,d\bar{M} + P_l(M). \tag{85}$$

Both the z and ρ equations are differential equation of the Hill type; ϕ_{rs} can be obtained from Eq. (84) by quadrature once ρ_{rs} is known.

In order to obtain solutions to the above equations as functions of the time, one must transform all the coefficients of these equations which depend on the true anomaly f of the elliptic motion into equivalent expressions of the mean anomaly M. Such a transformation is defined by Eq. (111) in Chapter II; its coefficients are given in Table 2.13 as polynomials in the orbital eccentricity e_0.

The solutions to the equations of the Hill type which were encountered above can be obtained in a recursive fashion as power series in e_0, each approximation depending on a linear differential equation with constant coefficients.

An extensive discussion of the feasibility of the general solution as well as explicit results concerning the first- and second-order approximations were provided by Lanzano (1969, 1971). The same author has also elaborated a procedure for eliminating the secular terms in the translational motion of the centroids that appear in higher-order approximations and that cannot be justified in a mathematical modeling of the Earth–Moon binary system.

Extensive analytic expansions are required to proceed to the third- and higher-order approximations; however, once these operations are programmed and can be routinely performed on a computer, this approach promises to provide an accurate solution of the perturbed motion of Earth's axis due to a real size oblate Moon.

We end this section by providing some information on the first-order approximation for the motion of Earth's rotational axis as obtained from the above formulation. In the first-order approximation, the nutational angle $\gamma_1(M)$ and precessional angle $\alpha_1(M)$ depend exclusively on Earth's oblateness and the zero-order approximation; the only lunar dependence is through the ratio $q_2 = m_2/(m_1 + m_2)$ of the Moon's mass to the total mass of the system. $\gamma_1(M)$ is obtainable through Eq. (82); $\alpha_1(M)$ and $\beta_1(M)$ are subsequently computed by quadratures. $\alpha_1(M)$ always has a secular term. The expressions for these two quantities are of the form

$$\begin{aligned} \alpha_1(M) &= S_\alpha M - \frac{E_0\Gamma_0}{\Lambda_0^3} + q_2 \cdot (\text{periodic terms}), \\ \gamma_1(M) &= \frac{\Gamma_0}{\Lambda_0^2}(1 - \cos\Lambda_0 M)\sin\gamma_0 + q_2 \cdot (\text{periodic terms}). \end{aligned} \tag{86}$$

The secular term is given by

$$S_\alpha = \frac{E_0\Gamma_0}{\Lambda_0^2} - \frac{3}{8}q_2\left[\frac{G_0E_0}{\Lambda_0^2}\cos\gamma_0 + \left(\frac{E_0^2}{\Lambda_0^2} - 1\right)Q\cos 2\psi\right]. \tag{87}$$

In the above equations we have written

$$
\begin{gathered}
E_0 = \beta'_0 - \alpha'_0 \cos\gamma_0, \qquad \Gamma_0 = -\tfrac{1}{2}\alpha'_0(\beta'_0 + \alpha'_0\cos\gamma_0), \\
\Lambda_0^2 = (\alpha'_0)^2 + E_0\beta'_0, \qquad q_2 = m_2/(m_1 + m_2), \qquad \psi = \omega_0 - \alpha_0, \\
G_0 = 2 + 3e_0^2, \qquad \sigma_r = r - 2\alpha'_0, \\
Q = -\left(\frac{e_0}{\sigma_1} + \frac{5e_0^2 - 2}{\sigma_2} - \frac{7e_0}{\sigma_3} - \frac{17e_0^2}{\sigma_4}\right),
\end{gathered}
\tag{88}
$$

and the values α'_0, β'_0, γ'_0 refer to Earth's axis.

The periodic terms appearing in the above equations have arguments of the type $\sigma_r M \pm 2\psi$ and $\Lambda_0 M \pm 2\psi$, where $\psi = \omega_0 - \alpha_0$ is the phase angle between the longitude of the centroid when it occupies the periastron position and the longitude of the nodal line. The coefficients of these periodic terms are rational functions of Λ_0. More details can be found in Lanzano (1969, 1971).

Chapter VI | Ocean Tides

6.00 INTRODUCTION

As an application of the elastic deformation of the Earth we shall discuss in this chapter the topic of the global ocean tides.

We begin by discussing the classical Laplace tidal equation (LTE) pertaining to an ocean covering a spherical, rotating Earth. We proceed with the study of some series solutions to that equation which are valid for the case of constant depth and which were obtained by Darwin and Hough.

After introducing the tidal potential due to the perturbations of the Moon and Sun, we give a short classification of the tides.

We proceed next to formulate a more realistic model by considering the effects of (1) the yielding of the solid Earth expressed by the load numbers; (2) friction at the bottom; and (3) the introduction of a continental boundary. We thus formulate a boundary-value problem governed by integrodifferential equations whose solutions must be obtained by computer.

We close the chapter by providing some information on the numerical procedures followed in recent work in this field.

6.01 LAPLACE TIDAL EQUATIONS

We consider the oscillations of a shallow ocean covering a rotating Earth. The equilibrium configuration of the composite system formed by the solid Earth and the fluid ocean will be an equipotential surface of the geopotential U (i.e., the potential of the gravitational and centrifugal forces).

The position of each particle within the fluid ocean can be determined by means of the colatitude θ, the eastward longitude ϕ, and the elevation z measured upward along the normal to the meridian curve.

The ocean depth $h(\theta, \phi)$ denotes the distance from the water surface to the ocean bottom averaged over a long period of time. We denote by ζ and δ the elevations of the ocean surface and ocean bottom above their average values. Thus the tide measured by conventional or deep-sea gauges fixed to Earth's surface is $\zeta_0 = \zeta - \delta$. The kinetic energy of the unit of mass of the fluid can be written as

$$2T = (R + z)^2\dot{\theta}^2 + (R + z)^2 \sin^2\theta\,(\omega + \dot{\phi})^2 + \dot{z}^2, \tag{1}$$

where R is the radius of curvature of the meridian and $\omega = 0.729211 \times 10^{-4}$ rad/sec is Earth's rotational velocity. Dots denote derivatives with respect to time. We denote by $V(\theta, \phi, z)$ the total potential energy and write $V = U + \Gamma$, where U is the geopotential and Γ is the potential of the disturbing forces. The latter includes not only the astronomical component Ω due to the Moon and the Sun obtainable from the motion of these perturbing bodies but also a contribution due to the yielding of the solid Earth.

The Lagrangian equations of motion,

$$\frac{d}{dt}\left(\frac{\partial T}{\partial \dot{q}}\right) - \frac{\partial T}{\partial q} = -\frac{\partial V}{\partial q},$$

yield the following system:

$$\begin{aligned}
\frac{d}{dt}[(R + z)^2\dot{\theta}] - (R + z)^2(\omega + \dot{\phi})^2 \sin\theta\cos\theta &= -\frac{\partial V}{\partial \theta},\\
\frac{d}{dt}[(R + z)^2 \sin^2\theta\,(\omega + \dot{\phi})] &= -\frac{\partial V}{\partial \phi},\\
\frac{d}{dt}(\dot{z}) - (R + z)\dot{\theta}^2 - (R + z)\sin^2\theta\,(\omega + \dot{\phi})^2 &= -\frac{\partial V}{\partial z}.
\end{aligned} \tag{2}$$

We introduce the following notation: $(R + z)\dot{\theta} = u$ for the southward component of the velocity (positive for increasing θ); $(R + z)\sin\theta\,\dot{\phi} = v$ for the eastward component of the velocity (positive for increasing ϕ toward the east); and $\dot{z} = w$ for the upward (vertical) component of the velocity. We expand Eqs. (2) and agree to neglect (1) the quadratic terms in u, v, w with respect to the corresponding linear terms; (2) the vertical component w with respect to the horizontal components u, v; (3) the derivative $\partial w/\partial t$ in comparison with the ωv term; (4) the water height z with respect to the radius of curvature R of the meridian curve; and (5) terms in $R\omega^2$. We can easily establish the following result:

$$\frac{\partial u}{\partial t} - 2\omega v \cos\theta = -\frac{1}{R}\frac{\partial V}{\partial \theta},$$
$$\frac{\partial v}{\partial t} + 2\omega u \cos\theta = -\frac{1}{R\sin\theta}\frac{\partial V}{\partial \phi}, \tag{3}$$
$$2\omega v \sin\theta = \frac{\partial V}{\partial z}.$$

We next integrate the last equation of the above set through the disturbed layer, i.e., between the limits $z = h$ and $z = h + \zeta$. Since $z = h$ represents the equilibrium configuration of the geopotential, we must have $U(h) = \text{const.}$; it follows that

$$U(h + \zeta) = U(h) + g\zeta = \text{const.} + g\zeta,$$

where $g = (\partial U/\partial z)_{(z=h)}$ is the gravity at the level surface $z = h$. The aforementioned integration yields

$$2\omega \sin\theta \int_h^{h+\zeta} v(z)\,dz = V(h + \zeta) - V(h) = g\zeta,$$

where it is assumed that $\Gamma(\theta, \phi, R + z)$ is constant across the height of the oceanic layer. The term on the left-hand side is of the order of $\omega v\zeta$ and can be dropped in comparison with $g\zeta$ since $\omega v \ll g$. We have then

$$\begin{aligned} V(h + \zeta) = V(h) + g\zeta &= U(h) + g(\zeta - \bar{\zeta}) \\ &= \text{const.} + g(\zeta - \bar{\zeta}), \end{aligned}$$

where $\bar{\zeta} = -\Gamma(h)/g(h)$.

We substitute this approximation into the first two equations of the set (3) and perform a final simplification by neglecting the ellipticity of the Earth. We can then take $R = a = 6371$ km as the mean radius, and having neglected terms in $a\omega^2$, we can assume g to be constant over the spherical surface, equal to the mean gravity $g = Gm_1/a^2 = 9.82024\ \text{m/sec}^2$. We finally get

$$\frac{\partial u}{\partial t} - 2\omega v \cos\theta = -\frac{g}{a}\frac{\partial}{\partial \theta}(\zeta - \bar{\zeta}),$$
$$\frac{\partial v}{\partial t} + 2\omega u \cos\theta = -\frac{g}{a\sin\theta}\frac{\partial}{\partial \phi}(\zeta - \bar{\zeta}). \tag{4}$$

To the above system, we can add a third equation which represents the condition of continuity. For this purpose, let us note that the flux of fluid exiting the volume of height $D = h + \zeta_0$ and area base $(a \sin\theta\, d\phi)(a\, d\theta)$ is represented by

$$\frac{\partial}{\partial\theta}(a\sin\theta\, d\phi\, Du)\, d\theta + \frac{\partial}{\partial\phi}(a\, d\theta\, Dv)\, d\phi$$

and that for reasons of continuity, it shall equal the change in volume due to the variation of the free surface $\zeta - \delta = \zeta_0$, which is represented by

$$-(a\sin\theta\, d\phi)(a\, d\theta)(\partial\zeta_0/\partial t).$$

By equating these two terms we obtain

$$\frac{\partial}{\partial\theta}(Du\sin\theta) + \frac{\partial}{\partial\phi}(Dv) = -a\sin\theta\frac{\partial\zeta_0}{\partial t}. \tag{5}$$

Equations (4) and (5) are known as the Laplace tidal equations (LTE) and refer to a spherical, rotating Earth. If no yielding of the Earth is taken into account, then $\delta = 0$ and $\zeta_0 = \zeta$. Also, in a first-order approximation, one can neglect ζ with respect to h and write h instead of D in Eq. (5).

6.02 DARWIN'S POWER SERIES SOLUTIONS

We shall consider some classical results first established by G. H. Darwin regarding the oscillations of an ocean layer covering the whole Earth and assumed to be (1) a surface of revolution, i.e., independent of the longitude ϕ, and (2) of constant depth h. If we assume the solutions to have the time factor $\exp(i\sigma t)$, then Eqs. (4) and (5) can be time differentiated and solved for u, v, and ζ. We introduce new variables

$$\mu = \cos\theta, \qquad \zeta^* = \zeta - \overline{\zeta} \tag{6}$$

and the two nondimensional constants

$$f = \sigma/2\omega, \qquad \beta = 4\omega^2 a^2/hg. \tag{7}$$

We obtain the system

$$\begin{aligned} hu &= \frac{i\sigma a}{\beta}\frac{(1-\mu^2)^{1/2}}{\mu^2 - f^2}\frac{\partial\zeta^*}{\partial\mu}, \\ hv &= \frac{2\omega a}{\beta}\frac{\mu(1-\mu^2)^{1/2}}{f^2-\mu^2}\frac{\partial\zeta^*}{\partial\mu}, \\ i\sigma a\zeta &= \frac{\partial}{\partial\mu}(hu(1-\mu^2)^{1/2}). \end{aligned} \tag{8}$$

Elimination of the u-variable between the first and third equations of this set yields

$$\frac{\partial}{\partial\mu}\left(\frac{1-\mu^2}{\mu^2-f^2}\frac{\partial\zeta^*}{\partial\mu}\right)=\beta(\zeta^*+\bar{\zeta}). \tag{9}$$

We look for a solution to Eq. (9) of the form

$$\frac{1}{\mu^2-f^2}\frac{\partial\zeta^*}{\partial\mu}=A_1\mu+A_3\mu^3+\cdots+A_{2j+1}\mu^{2j+1}+\cdots, \tag{10}$$

with coefficients to be determined. This series is applicable to the case of symmetry with respect to the equatorial plane because by integration it leads to the expression

$$\zeta^*=C-\frac{1}{2}f^2A_1\mu^2+\frac{1}{4}(A_1-f^2A_3)\mu^4+\cdots+\frac{1}{2j}(A_{2j-3}-f^2A_{2j-1})\mu^{2j}+\cdots, \tag{11}$$

which contains only even powers of μ. Here C is an arbitrary constant. In terms of the trial series given by Eq. (10), the left-hand side of Eq. (9) becomes

$$\frac{\partial}{\partial\mu}\left(\frac{1-\mu^2}{\mu^2-f^2}\frac{\partial\zeta^*}{\partial\mu}\right)=A_1+3(A_3-A_1)\mu^2$$
$$+\cdots+(2j+1)(A_{2j+1}-A_{2j-1})\mu^{2j}+\cdots. \tag{12}$$

We consider next a perturbing potential $\Gamma=\Omega=-g\bar{\zeta}$, where

$$\bar{\zeta}=H_1(\tfrac{1}{3}-\mu^2)\cos(\sigma t), \tag{13}$$

which consists only of the astronomical contribution and includes the lunar fortnightly and the solar semiannual tides for appropriate values of σ. More particularly, should the constant H_1 vanish, there would be free oscillations; in this case, however, the spectrum of allowable frequencies σ must be determined.

By substituting Eqs. (11)–(13) in (9) and equating the coefficients of like powers, we get the following recurrent relations:

$$\begin{aligned} A_1&=\beta(C+\tfrac{1}{3}H_1),\\ A_{2j+1}-(1-f^2B_j)A_{2j-1}-B_jA_{2j-3}&=0, \qquad j\geqslant 1, \end{aligned} \tag{14}$$

where

$$B_j=\beta/2j(2j+1), \tag{15}$$

with the understanding that $A_{-1}=-2H_1$.

Let us briefly discuss the preceding three-term relation and ascertain under what conditions, or equivalently for which values of f or β, it determines a series which converges everywhere including at the poles ($\mu = \pm 1$). For only then can there be finite velocities everywhere and both $u = 0$ and $v = 0$ at the poles, as required from Eqs. (10) and (8).

The vanishing of u and v at the poles is justified from the point of view of global symmetry. We can also interpret the vanishing of u by a limiting procedure. If there are boundaries to the ocean layer along parallels of latitude, then u should vanish on each of these boundaries. Should these boundaries recede to the poles, so that an ocean layer covers the whole sphere, u must still be zero for $\mu = \pm 1$.

We introduce the notation

$$N_{j+1} = A_{2j+1}/A_{2j-1}, \tag{16}$$

and rewrite Eq. (14) for $j \geqslant 1$ as

$$[N_{j+1} - (1 - f^2 B_j)]N_j = B_j, \tag{17}$$

and note that the B_j constitute a null sequence; i.e., they tend to zero for j increasing to infinity. This shows that the limit N of N_j as $j \to \infty$ must satisfy the condition

$$(N - 1)N = 0;$$

we must therefore have either $N = 0$ or $N = 1$.

When $N = 0$, the well-known ratio test for the convergence of infinite series shows that both series (10) and (11) are convergent everywhere, including at the poles ($\mu = \pm 1$). When $N = 1$, we first note that from a certain point onward the A coefficients will have the same sign. We make use of the less-known Raabe convergence test (see Knopp, 1951, p. 285). This test applies to series with elements a_n of the same sign and states that if

$$\lim_{n \to \infty} n\left(\frac{a_{n+1}}{a_n} - 1\right)$$

is > -1, then the series is divergent, and that if this limit is $\leqslant -1$, then the series is convergent. From Eqs. (16) and (17), we can make the approximation

$$N_{j+1} = \frac{A_{2j+1}}{A_{2j-1}} \cong 1 - (f^2 - 1)B_j$$

and evaluate the requisite limit for the expression

$$(2j - 1)\left(\frac{A_{2j+1}}{A_{2j-1}} - 1\right) = -\frac{(f^2 - 1)(2j - 1)\beta}{2j(2j + 1)}.$$

Since it approaches zero as $j \to \infty$, we can state that the series (10) is divergent for $\mu = \pm 1$. This situation is not acceptable from a physical point of view. We have therefore established the result that in order to reach a plausible solution for an ocean covering the globe, we must have $N = 0$.

From Eq. (17) we have

$$N_j = \frac{-B_j}{(1 - f^2 B_j) - N_{j+1}}. \tag{18}$$

Since N_j must approach zero, repeated application of this expression will yield a convergent continued fraction

$$N_j = -\frac{B_j}{1 - f^2 B_j +}\,\frac{B_{j+1}}{1 - f^2 B_{j+1} +}\,\frac{B_{j+2}}{1 - f^2 B_{j+2} +}\cdots. \tag{19}$$

The truncation of this infinite expansion at the term corresponding to $j + n$ gives rise to a rational expression called the convergent of the continued fraction; we shall denote it by C_{j+n}/D_{j+n} and note that it can most expeditiously be evaluated by means of the matrix multiplication scheme

$$\begin{bmatrix} C_{j+n} \\ D_{j+n} \end{bmatrix} = \prod_{k=j}^{j+n-1} \begin{bmatrix} 0 & B_k \\ 1 & 1 - f^2 B_k \end{bmatrix} \begin{bmatrix} B_{j+n} \\ 1 - f^2 B_{j+n} \end{bmatrix}.$$

We consider first the case $H_1 \neq 0$, so that the value of the frequency σ is known. Equation (19) determines N_1, so that we can write

$$A_1 = N_1 A_{-1} = -2N_1 H_1 \tag{20}$$

because of Eq. (16) and the definition of A_{-1}. Using the first of Eq. (14) and this value of A_1 we can solve for the constant, which up to now has been arbitrary, to obtain

$$C/H_1 = -\tfrac{1}{3} - (2/\beta)N_1. \tag{21}$$

This value of C is the only one consistent with a zero limit for N_j and a zero velocity at the poles.

The recurrent relation (14) gives rise to

$$\begin{gathered} A_1/H_1 = -2N_1, \qquad A_3/H_1 = N_2 A_1/H_1 = -2N_1 N_2, \\ A_5/H_1 = N_3 A_3/H_1 = -2N_1 N_2 N_3, \\ A_{2j+1}/H_1 = N_{j+1} A_{2j-1}/H_1 = -2N_1 N_2 \cdots N_{j+1}. \end{gathered} \tag{22}$$

Using Eqs. (11), (13), (21), and (22), we can finally evaluate the tidal elevation:

$$\begin{aligned} \frac{\zeta}{H_1} = \frac{\zeta^* + \bar{\zeta}}{H_1} &= -\frac{2}{\beta} N_1 - (1 - f^2 N_1)\mu^2 \\ &\quad - \frac{1}{2} N_1 (1 - f^2 N_2)\mu^4 + \cdots - \frac{1}{j}(N_1 N_2 \cdots N_{j-1})(1 - f^2 N_j)\mu^{2j} + \cdots. \end{aligned} \tag{23}$$

The most convenient procedure is to evaluate N_j for a large value of j as the convergent of a continued fraction by means of the matrix multiplication procedure, and then obtain $N_{j-1}, N_{j-2}, \ldots, N_1$ by using the fundamental equation (18).

The case of free oscillations, i.e., $H_1 = 0$, is different. One has to find the values of σ which cause N_j to tend to zero. We can get such a condition by equating the value of N_2 given by Eq. (19) with the value $N_2 = 1 - B_1 f^2$ obtainable from Eq. (14). We get

$$1 - B_1 f^2 = \frac{-B_2}{1 - f^2 B_2 +} \frac{B_3}{1 - f^2 B_3 +} \cdots \cong \frac{C_{2+n}}{D_{2+n}}, \tag{24}$$

where we have approximated the continued fraction by its convergent of appropriate dimension. From Eq. (24) we can determine the admissible values for f or σ. Once the values of f are known, we can solve for $A_1, A_3, \ldots,$ where C is left arbitrary:

$$A_1 = \beta C, \qquad A_3 = N_2 A_1 = N_2 \beta C, \ldots .$$

For the original work, see Darwin (1886, pp. 337–342); also (1907, pp. 366–371).

6.03 HOUGH'S SOLUTIONS IN SPHERICAL HARMONICS

We continue the study of Eq. (9), which is valid for a uniform ocean depth, and assume that the tidal elevation ζ is expressible as a series of Legendre polynomials

$$\zeta = \sum_{n=0}^{\infty} C_n P_n(\mu), \tag{25}$$

where $\mu = \cos\theta$.

In considering the disturbing potential, we wish to take into account the attraction that the layer of tidal water (of height ζ) exerts on the rest of the Earth, including its oceanic mass, considered as a sphere of mean radius a and mean density $\bar{\rho}_0$. The mass of the tidal water is

$$m_{\mathrm{w}} = 4\pi a^2 \zeta \rho_{\mathrm{w}},$$

where ρ_{w} stands for the density of the ocean water. The expressions for the outer and inner potentials due to this mass are

$$\frac{G m_{\mathrm{w}}}{2n+1} \frac{a^n}{r^{n+1}} \quad \text{and} \quad \frac{G m_{\mathrm{w}}}{2n+1} \frac{r^n}{a^{n+1}};$$

respectively. These two expressions coincide for $r = a$, giving rise to

$$\frac{4\pi Ga}{2n+1}\rho_w\zeta.$$

We eliminate the factor $4\pi Ga$ from $3g(a) = 4\pi Ga\bar{\rho}_0$ and represent the potential of the tidal layer as

$$\frac{3g\rho_w\zeta}{(2n+1)\bar{\rho}_0},$$

where both $\bar{\rho}_0$ and g are to be evaluated at $r = a$. We include this tidal contribution within the disturbing potential and write

$$\Omega = -g\sum_{n=0}^{\infty}\left(\bar{\zeta}_n + \frac{3}{2n+1}\frac{\rho_w}{\bar{\rho}_0}\zeta_n\right),$$

so that

$$\zeta^* = \sum_{n=0}^{\infty}(\alpha_n\zeta_n - \bar{\zeta}_n), \tag{26}$$

where

$$\alpha_n = 1 - \frac{3}{2n+1}\frac{\rho_w}{\bar{\rho}_0} \tag{27}$$

is a dimensionless quantity, with $\rho_w/\bar{\rho}_0 \cong 0.181$. If we write

$$\bar{\zeta} = \sum_{n=0}^{\infty}\gamma_n P_n(\mu), \tag{28}$$

then

$$\zeta^* = \sum_{n=0}^{\infty}(\alpha_n C_n - \gamma_n)P_n(\mu). \tag{29}$$

We next substitute Eqs. (28) and (29) in Eq. (9) and integrate:

$$\sum_{n=0}^{\infty}(\alpha_n C_n - \gamma_n)(1-\mu^2)\frac{dP_n}{d\mu} = \beta[(1-f^2)-(1-\mu^2)]\sum_{n=0}^{\infty}C_n\int_{-1}^{\mu}P_n(\mu)\,d\mu. \tag{30}$$

No integration constant is required here because both sides vanish for $\mu = -1$.

To extract recurrent relations from Eq. (30), we must represent both integral expressions appearing in its right-hand side as a sum of the differential terms appearing in the left-hand side. For this purpose we make use of the fundamental relation valid for the Legendre polynomials and of formulas (36) and (34) listed in Hobson's treatise (1955, p. 33). We have

$$\int_{-1}^{\mu} P_n(\mu)\,d\mu = -\frac{1-\mu^2}{n(n+1)}\frac{dP_n}{d\mu} = \frac{1}{(2n+1)}(P_{n+1} - P_{n-1})$$

$$= \frac{1}{(2n+1)}\left[\frac{1}{(2n+3)}\frac{dP_{n+2}}{d\mu} - 2\frac{(2n+1)}{(2n+3)(2n-1)}\frac{dP_n}{d\mu} + \frac{1}{(2n-1)}\frac{dP_{n-2}}{d\mu}\right].$$

We use the second term appearing in the above formula to represent the first integral term, the one multiplied by $(1 - f^2)$; we use the last term of the above formula to eliminate the integral term which is multiplied by $(1 - \mu^2)$. By doing so, we reach the following recurrent relation:

$$\frac{C_{n-2}}{(2n-3)(2n-1)} - C_n L_n + \frac{C_{n+2}}{(2n+3)(2n+5)} = \frac{\gamma_n}{\beta}, \tag{31}$$

where

$$L_n = \frac{f^2 - 1}{n(n+1)} + \frac{2}{(2n-1)(2n+3)} - \frac{\alpha_n}{\beta}, \tag{32}$$

This is valid for $n \geqslant 1$, provided that we consider $C_0 = C_{-1} = 0$.

It is clear that Eq. (31) preserves indicial parity. We shall proceed to treat the case of even-order subscripts, which represents tides symmetric with respect to the equatorial plane. Similar procedures apply for the case of unsymmetric tides.

Let us consider first the case of free oscillations, for which $\gamma_n = 0$. Equation (31) will then represent an infinite system of homogeneous linear equations in $C_2, C_4, \ldots, C_n, \ldots$ for even n. For the convergence of the ζ-series it is necessary, although by no means sufficient, that the limit of C_n as $n \to \infty$ be zero. In practice the system can be terminated with an appropriate C_ν, assuming that the C with higher subscripts are negligible. The consistency of this infinite system can be expressed by the vanishing of the determinant formed with the coefficients of the unknowns. This determinant is easily seen to consist of the main diagonal $(-L_2, -L_4, \ldots, -L_n, \ldots)$ bordered above by the upper diagonal sequence $1/(2n+3)(2n+5)$ and below by the lower diagonal sequence $1/(2n-3)(2n-1)$. The dominant term in the expansion of this determinant is the main diagonal, so that in a first-order approximation, the roots of $L_2 = 0, L_4 = 0, \ldots, L_n = 0, \ldots$ should provide the values of the frequencies that cause the system to be consistent.

To provide a more efficient algorithm for the evaluation of the frequencies of free oscillations, we let

$$\epsilon_n = \frac{1}{(2n+1)(2n+3)^2(2n+5)} \tag{33}$$

and introduce the following two continued fractions:

$$H_n = \frac{\epsilon_n|}{|L_n} - \frac{\epsilon_{n-2}|}{|L_{n-2}} - \cdots \frac{\epsilon_2|}{|L_2},$$

which terminates at H_2, and the infinite continued fraction

$$K_n = \frac{\epsilon_{n-2}|}{|L_n} - \frac{\epsilon_n|}{|L_{n+2}} - \cdots.$$

They satisfy the recurrent formulas

$$H_n = \frac{\epsilon_n}{L_n - H_{n-2}}, \qquad K_n = \frac{\epsilon_{n-2}}{L_n - K_{n+2}}. \tag{34}$$

We can then rewrite Eq. (31) as

$$\frac{C_{n-2}/C_n}{(2n-3)(2n-1)} = L_n - \frac{1/[(2n+3)(2n+5)]}{C_n/C_{n+2}}$$

$$= L_n - \frac{\epsilon_n}{(C_n/C_{n+2})/[(2n+1)(2n+3)]}.$$

By repeated application of this formula and using Eq. (34), we obtain the following result:

$$\begin{aligned} C_{n-2}/C_n &= (2n-3)(2n-1)(L_n - K_{n+2}) \\ &= (2n-3)(2n-1)\epsilon_{n-2}/K_n = 1/(2n-1)(2n+1)K_n, \end{aligned}$$

or

$$C_n/C_{n-2} = (2n-1)(2n+1)K_n. \tag{35}$$

In a similar fashion we obtain

$$\begin{aligned} \frac{C_{n+2}/C_n}{(2n+3)(2n+5)} &= L_n - \frac{\epsilon_{n-2}}{(C_n/C_{n-2})/[(2n-1)(2n+1)]} \\ &= L_n - H_{n-2} = \epsilon_n/H_n, \end{aligned} \tag{36}$$

and this leads to

$$C_n/C_{n+2} = H_n/(2n+3)(2n+5)\epsilon_n = (2n+1)(2n+3)H_n. \tag{37}$$

Replacing the appropriate version of Eq. (35) in the left-hand side of Eq. (36) and equating this to the third term, we have the important and yet simple relation

$$K_{n+2} = L_n - H_{n-2}, \tag{38}$$

valid for any $n \geqslant 2$. This equation represents the consistency condition of the infinite homogeneous linear system satisfied by the C. There exists an infinite number of such equations, each of which is equivalent to the original determinant equation. However, each equation is written down (for a given n) in a form that is best suited to approximate that root of the equation which differs slightly from the root of $L_n = 0$.

Thus in the first-order approximation, the frequencies of free oscillation are given by

$$\frac{\sigma_n^2}{4\omega^2} = 1 + n(n+1)\left[\frac{hg\alpha_n}{4\omega^2 a^2} - \frac{2}{(2n-1)(2n+3)}\right]. \tag{39}$$

They can be improved upon by using Eq. (38), where the various convergents of the two continued fractions can be most expeditiously calculated from Eq. (34). The frequencies of free oscillation for a nonrotating ocean of the same depth are known to be given by the sequence

$$\sigma_n^2 = n(n+1)\frac{hg\alpha_n}{a^2},$$

and the agreement between the two cases improves with increasing n.

Corresponding to a value of σ_n, one can evaluate the tidal elevation by repeated use of Eqs. (35) and (37). Thus for even values of both n and k, we obtain

$$\begin{aligned}\zeta = \exp(i\sigma_n t)\,[&C_n P_n + C_{n+2}P_{n+2} + \cdots + C_{n+k}P_{n+k} \\ &+ \cdots + C_{n-2}P_{n-2} + C_{n-4}P_{n-4} + \cdots + C_0 P_0],\end{aligned} \tag{40}$$

where

$$\begin{aligned}C_{n+k} &= (2n+3)(2n+5)\cdots(2n+2k+1)K_{n+2}K_{n+4}\cdots K_{n+k}C_n, \\ C_{n-k} &= (2n-1)(2n-3)\cdots(2n-2k+1)H_{n-2}H_{n-4}\cdots H_{n-k}C_n,\end{aligned} \tag{41}$$

and C_n remains arbitrary.

The study of forced oscillations follows immediately from the preceding considerations. We shall limit ourselves to a disturbing potential expressible by means of a single harmonic $\gamma_r P_r(\mu)\exp(i\sigma_r^* t)$ with given γ and σ^*. The presence of more harmonics can be dealt with by superimposing the results. The C will now satisfy a nonhomogeneous system because the rth equation is

$$C_{r-2}/(2r-3)(2r-1) - C_r L_r + C_{r+2}/(2r+3)(2r+5) = \gamma_r/\beta. \tag{42}$$

The vanishing of the C for $n \to \infty$ is still a valid assumption for convergence. Equation (35) will still be valid for $n > r$, and Eq. (37) for $n < r$; this will be so, independently of any consistency considerations, because they simply represent the ratio of the two variables for decreasing (respectively, increasing) values of the subscripts. Substituting these ratios in Eq. (42), we find

$$C_r(H_{r-2} - L_r + K_{r+2}) = \gamma_r/\beta, \tag{43}$$

which determines the value of C_r. The given frequency σ_r^* should not coincide with the frequency σ_r of free oscillation because otherwise the expression in parentheses will vanish and it will be impossible to determine C_r. We next represent the tidal elevation according to Eqs. (40) and (41); in this case, however, the value of C_r must be provided by Eq. (43).

Fundamental references here are Hough (1897, 1898). A summary of these results appears also in Darwin (1911).

The LTE and some of these classical series solutions are also discussed in Lamb (1945, pp. 330–348).

6.04 POTENTIAL OF THE TIDE-GENERATING FORCES

The gravitational attraction exerted by the Moon upon any point Q located on the Earth's surface at a distance r from its center can be expressed as

$$-Gm/(d^2 - 2rd\cos\gamma + r^2)^{1/2}.$$

Here G is the universal gravitational constant, m the Moon's mass, d the distance between the Moon's center and the Earth's center, and γ the Moon's zenithal distance from Q, that is to say, the angular distance between Q and the Moon's center measured at the Earth's center.

By tidal motion at Q generated by the Moon, we understand the gravitational acceleration at Q due to the Moon measured with respect to the gravitational acceleration experienced by the Earth's center because of the Moon's attraction.

The acceleration of the Earth's center amounts to Gm/d^2 acting in the direction of the line joining the two centers. The potential due to this uniform field is $-Gmr\cos\gamma/d^2$. The potential of the relative attraction at Q is then given by

$$\Omega = -Gm(d^2 - 2rd\cos\gamma + r^2)^{-1/2} + Gmr\cos\gamma/d^2.$$

We expand the above expression into a power series of the small parameter r/d; the leading terms appearing in this expansion are seen to be

$$\Omega = -Gm/d - Gmr^2P_2(\cos\gamma)/d^3 - \cdots = \Omega_0 + \Omega_2 + \cdots. \tag{44}$$

We consider next the equilibrium configuration of a self-gravitating, rotating Earth given by $U(h, \theta, \phi) = \text{const.}$, where U is the potential of the self-gravitation and centrifugal forces. The height $\bar{\zeta}$ due to the tidal perturbations, when measured above the equilibrium height h, can be obtained from the equipotential condition

$$U(h + \bar{\zeta}) + \Omega_2(h + \bar{\zeta}) = \text{const.}$$

This tidal spheroid can be approximated by the equation

$$U(h) + \bar{\zeta} g(h) + \Omega_2(h) = \text{const.}, \tag{45}$$

where $g(h)$ is the value of $\partial U/\partial z$ at $z = h$ and we have assumed that Ω_2 has no appreciable variation in the interval being considered. For a spherical Earth and with the order of approximation already assumed, we have $h = r = a$ and $g(a) = Gm_1/a^2$.

Because $U(h)$ is a constant, we get from Eq. (45)

$$\begin{aligned} \bar{\zeta} &= -\Omega_2(a)/g(a) + \text{const.} \\ &= H(c/d)^3(\cos^2 \gamma - \tfrac{1}{3}) + \text{const.}, \end{aligned} \tag{46}$$

where

$$H = \frac{3}{2} \frac{m}{m_1} (a/c)^3 a \cong 53.7 \text{ cm}, \tag{47}$$

and it is assumed that $m_1/m = 81.31$, $a = 6{,}371$ km, and $c = 384{,}400$ km for the mean Earth–Moon distance.

A simple computation will show that in the case of solar tides the corresponding constant H' will be related to the lunar tidal constant H by

$$H' = 0.45990H. \tag{48}$$

Because of the Earth's diurnal rotation and the Moon's orbital motion, the shape of the tidal spheroid will change with time; this is equivalent to saying that for any fixed position on Earth the level of the ocean surface will rise and fall.

To express this variation analytically, let us denote by δ the Moon's angular distance from the North Pole and by α the Moon's hour angle measured westward from a fixed meridian. Assuming, as is customary, that the longitude ϕ of Q is measured eastward from the same meridian, we can write

$$\cos \gamma = \cos \delta \cos \theta + \sin \delta \sin \theta \cos(\phi + \alpha),$$

where θ is the colatitude of Q.

Substituting this expression in Eq. (46), after a few trigonometrical manipulations we get

$$\begin{aligned}\bar{\zeta} = \tfrac{1}{2}H(c/d)^3[&3(\cos^2\delta - \tfrac{1}{3})(\cos^2\theta - \tfrac{1}{3})\\ &+ \sin 2\delta \sin 2\theta \cos(\alpha + \phi)\\ &+ \sin^2\delta \sin^2\theta \cos(2\alpha + 2\phi)] + \text{const.}\end{aligned} \tag{49}$$

The three terms in the square brackets represent three additive partial tides, the zonal, tesseral, and sectorial types.

The first term depends on the colatitude of the given location Q only and vanishes along those parallels for which $\cos^2\theta = 1/3$. The tidal height varies as $\cos^2\delta - 1/3$ and exhibits a period which is equal to one half of the orbital period of the disturbing body. These are the lunar fortnightly and the solar semiannual tides. They are long-period tides, also known as declinational tides. Also, the mean value of the tidal elevation, averaged along the orbital path of the perturbing body, is not zero.

The second term represents a tidal height which depends on the hour angle of the disturbing body also. It vanishes at those meridians which are at 90° from the disturbing body and at the equator. It reaches a maximum height on the meridian of the disturbing body at 45° from the equator. The period depends on α; these are the diurnal tides.

The third term depends also on α and has a period of half a lunar or solar day; these are the semidiurnal tides. It vanishes along those meridians at 45° from the disturbing body.

Because of the invariance of the volume occupied by the oceanic mass, we must have

$$\iint \bar{\zeta}\, d\Sigma = 0, \tag{50}$$

the integration being extended over the surface of the whole ocean. This condition allows the determination of the constant term appearing in both Eqs. (46) and (49).

Suppose that the ocean covers the whole Earth; then we know that the integral of a spherical harmonic extended over the whole sphere vanishes, and so will the value of the constant.

In such circumstances, Eq. (46) reveals that the maximum tidal elevation will occur for $\gamma = 0$ and π, the minimum for $\gamma = \pi/2$, i.e., for the zenith and nadir positions of the perturbing body and when it lies on the location's horizon.

If we consider a bounded ocean, then Eq. (50) expresses the value of the parameter for each position of the disturbing body as a sum of spherical harmonics of the angles δ and α, the coefficients being integrals extended over the surface of the true ocean. Thus this constant quantity in reality varies with the position of the disturbing body and does influence the location and value of high tide: these values do not coincide any longer with the maxima exhibited by the disturbing potential.

6.05 SOME CONSIDERATIONS ON THE LUNAR ORBIT

The three quantities δ, α, and c/d which appear in the previous section exhibit considerable variations with time. The study of these variations is a problem of celestial mechanics and astronomy; it is essentially related to Newcomb's elements of the Earth's orbit (1960), which represent the apparent motion of the Sun, and to Brown's lunar theory (1905), which describes the lunar orbit.

A first attempt at providing a harmonic expansion of the tidal potential was due to Ferrel (1874) and was limited to only a few constituents. Darwin in 1883 published a more comprehensive development, which was extensively used for many years. His was not, however, a true harmonic expansion because he chose the lunar orbit rather than the ecliptic as the reference orbit; that caused some of the constants appearing in the coefficients and arguments to be slowly varying functions of time (see Darwin, 1907, pp. 1–69). Recently, Munk and Cartwright (1966) obtained a new development, limited to 13 harmonics, which is based on the Pontécoulant third-order theory of the lunar motion (see, e.g., Smart, 1953, p. 281). This theory contains terms up to the third power of the lunar eccentricity and of the ratio of the solar to lunar mean motions. The expansion is correct up to 10^{-2} rad.

In 1921 Doodson provided the most comprehensive expansion of the tidal potential by analytic manipulation of Brown's series of lunar theory. Brown's series express the coordinates of the Moon as the sum of about 1500 harmonic terms plus few secular effects. In order to facilitate the computations, he published the well-known tables (1919) of motion. However, in the construction of these tables he made approximations that reduced the accuracy of his theory. Since 1923 the Moon's orbit has been calculated using Brown's theory; the computations appearing in the *Astronomical Almanac* are based upon Brown's series rather than his tables.

Doodson's results reach up to terms having amplitudes of the order of 10^{-4} rad; it contains about 400 terms.

Eckert *et al.* (1954) reworked Brown's theory using a computer and introduced a few revisions into Brown's tables; this is known as the "Improved Lunar Ephemeris 1952–59"; it attains an accuracy of 10^{-7} rad.

In 1971 Cartwright and Tayler used this improved ephemeris to recheck Doodson's results. Using computer programs for tidal analysis, they generated time series from Moon and Sun ephemerides. Three series, each 18 years long, were analyzed, covering the periods 1861–1879, 1915–1933, and 1951–1969. Advanced filtering techniques were used to isolate the various constituents. Very few discrepancies with Doodson's results were discovered. Their analysis constitutes a definitive confirmation of the correctness of Doodson's tables, which had been obtained by hand computation. Doodson expressed Ω_2 and Ω_3 as series of tidal components according to the scheme

$$\Omega_n(\theta, \phi; t) = \sum \beta(n_1, n_2, \ldots, n_6) \times \exp[i(n_1\tau + n_2 s + n_3 h + n_4 p + n_5 N_1 + n_6 p_s)] \quad (51)$$

for $n = 2, 3$, where the summation is extended over all positive and negative values of the integers $n_1, n_2, \ldots, n_6$. The six arguments of the exponential $(\tau, s, h, p, N_1, p_s)$ are astronomical frequencies which will be described; they were so chosen because they happen to be practically linear functions of time over an interval of a century.

The set of integers

$$(n_1; n_2 + 5; n_3 + 5; n_4 + 5; n_5 + 5; n_6 + 5) \quad (52)$$

in this specified sequence is known as the argument number of the tide (also Doodson set) and corresponds uniquely to each individual component. The frequency of each component is

$$\sigma = n_1\dot{\tau} + n_2\dot{s} + \cdots + n_6\dot{p}_s, \quad (53)$$

the rate of change being usually per mean solar hour.

The six adopted variables are as follows:

1. τ is the mean lunar time (Moon's hour angle), measured from the passage of the Moon at the meridian;
2. s is the mean longitude of the Moon;
3. h is the mean longitude of the Sun;
4. p is the mean longitude of the lunar perigee;
5. N is the mean longitude of the mean lunar ascending node; $N_1 = -N$ is the quantity used in the development because N decreases with time due to the node's retrograde motion on the ecliptic;
6. p_s is the mean longitude of the perihelion.

The solar variables $(h; p_s)$ are measured along the ecliptic in the direction of motion from the mean equinox of date; the lunar variables (s, p, N) start similarly and follow the ecliptic as far as the mean lunar ascending node and then are measured along the lunar orbit.

The relation between τ, s, h and the mean solar time t (i.e., the Sun's hour angle) is easily seen to be

$$t + h = \tau + s = \theta, \quad (54)$$

in which we use the absolute values of these angles and recall that the longitude is reckoned eastward from the vernal equinox and that the hour angle is measured westward from the meridian. Here θ is the sidereal time, i.e., the hour angle of the vernal equinox.

τ has a period of 24 h and 50.47 min corresponding to the fact that the Moon arrives at the meridian about 50 min later every day; s has a period of 27.321 days, which is the time required for the Moon to return to the same declination; and h has a period of 365.242 days. p has a period of 3232 days, N_1 a period of 6798 days, and p_s a period of 20,940 years. The variables s, h, p, N_1, p_s can be expanded as polynomials of the form

$$A_0 + A_1 T + A_2 T^2 + A_3 T^3,$$

where T denotes Julian centuries of 36,525 ephemeris days counted from the epoch (i.e., the origin of time), which is taken to be 1900 January 0.5 ET (ephemeris time), i.e. Greenwich mean noon, and which corresponds to Julian date 2,415,020.0. The values of the coefficients A for the various variables are available in the *Astronomical Almanac.*

Besides these secular terms, these variables also have periodic terms; these can be obtained from Brown's series or tables; some of them have also been tabulated by Woolard (1953). The arguments used by both Brown and Woolard are as follows: $l = s - p$, the mean anomaly of the Moon; $l' = h - p_s$, the mean anomaly of the Sun; $F = s - N$, the mean angular distance of the Moon from its ascending node, also referred to as the Moon's argument of latitude; and $D = s - h$, the mean elongation of the Moon from the Sun.

Let us briefly examine the variations of the functions $(c/d)^3$, $\sin^2 \delta$, $\cos^2 \delta$, $\sin 2\delta$ which appear in Eq. (49) for the case of the Moon by using some of Brown's and Woolard's results. In this connection, let us remark that the hour angle α appearing in this equation is in effect the lunar time τ.

The quantity c/d is obtainable from the sine of the parallax, which in Brown's theory is a series containing more than 200 trigonometric terms of the form $\cos(l, l', F, D)$. Most of the older publications express these coefficients in seconds of arc; they can be reduced to radians by noting that

$$1'' = 4.848 \times 10^{-6} \text{ rad}$$

(i.e., $1'' \sim 5 \cdot 10^{-6}$ rad). Table 11 in Woolard's publication (1953, pp. 72–73) provides more than 100 terms of the expansion for $(c/d)^3$. Some of them are perturbations due to Venus and Jupiter; some of them are mixed periodic, i.e., of the form $T\cos(\ )$. The most important of these terms are the following:

$$\begin{aligned}(c/d)^3 = {} & 1.004739 + 0.164396 \cos l \\ & + 0.031475 \cos(l - 2D) + 0.026581 \cos 2D \\ & + 0.013443 \cos 2l + 0.001771 \cos(l' - 2D) \\ & + 0.001013 \cos(l - l') + 0.004188 \cos(l + 2D) \\ & + 0.001350 \cos(l - l' - 2D) + 0.001466 \cos 2(l - D) \\ & + 0.001077 \cos 3l + \cdots .\end{aligned} \tag{55}$$

The second term appearing in the right-hand side is due to the ellipticity of the orbit $e = 0.05490$; the third one is due to the evection; and the fourth one is due to the variation of the lunar orbit. The periods are, respectively, 27.554 days, 31.812 days, and one half of the synodic period (14.765 days). The ecliptic longitude of the Moon v is also tabulated by Woolard (1953, pp. 52–53), to more than 100 terms:

$$\begin{aligned} v = s &+ 0.109760 \sin l + 0.022236 \sin(l - 2D) \\ &+ 0.0011490 \sin 2D - 0.001996 \sin 2F \\ &- 0.003238 \sin l' - 0.001026 \sin 2(l - D) \\ &+ 0.003728 \sin 2l + \cdots . \end{aligned} \tag{56}$$

By solving a spherical triangle we get

$$\cos \delta = \sin \epsilon \sin v = 0.39782 \sin v,$$

where ϵ is the obliquity of the ecliptic, given by

$$\epsilon = 23^\circ\, 27'\, 8''.26 - 46''.845\, T,$$

and which for 1981 amounts to $\epsilon \cong 23^\circ.442$.

From this we get

$$\begin{aligned} \cos^2 \delta &= 0.079130(1 - \cos 2v), \\ \sin^2 \delta &= 0.920870 + 0.079130 \cos 2v. \end{aligned}$$

Since

$$\cos 2v = \cos(2s + 2\Delta) \cong \cos 2s - (2\Delta) \sin 2s,$$

where $\Delta = v - s$ is the infinite trigonometric series given Eq. (56), we can express both $\cos^2 \delta$ and $\sin^2 \delta$ as a sequence of terms containing the cosine of arguments obtained by adding and subtracting l, $2l$, l', $l - 2D$, $2(l - D)$, $2D$, $2F$, to $2s$.

An independent evaluation of the trigonometric functions of δ is obtainable by using the expression for the Moon's latitude β in tabular form (Woolard, 1953, p. 58),

$$\begin{aligned} \beta = 0.089503 \sin F &- 0.003023 \sin(F - 2D) \\ &+ 0.004897 \sin(l + F) + 0.004847 \sin(l - F) + \cdots , \end{aligned} \tag{57}$$

because then we can approximate as follows:

$$\begin{aligned} \cos^2 \delta &= \sin^2 \beta = \beta^2 - \tfrac{1}{3}\beta^4 + \cdots , \\ \sin^2 \delta &= \cos^2 \beta = 1 - \beta^2 + \tfrac{1}{3}\beta^4 + \cdots , \\ \sin \delta \cos \delta &= \cos \beta \sin \beta = \beta - \tfrac{2}{3}\beta^3 + \cdots . \end{aligned}$$

The above considerations provide all the ingredients for a direct evaluation of the three types of tides discussed in the previous section.

6.06 CLASSIFICATION OF TIDES

Using the leading terms appearing in the expansions for the lunar orbit, we can enumerate the lunar tides exhibiting the highest amplitudes.

A similar classification can be made for the solar tides by using the fact that the Earth–Sun distance d' can be represented as

$$(c'/d')^3 = 1 + 0.050 \cos l' + \cdots, \tag{58}$$

where the evection and variation terms are missing and the ellipticity term has a smaller coefficient than its lunar counterpart because the eccentricity of Earth's orbit is only $e' = 0.0167$.

To represent the sectorial tides of lunar origin, we must take into account the product

$$\begin{aligned}&[1.005 + 0.164 \cos l + 0.031 \cos(l - 2D) + 0.026 \cos 2D + \cdots] \\ &\quad \times (0.921 + 0.079 \cos 2s + \cdots) \cos 2\tau,\end{aligned} \tag{59}$$

where only the leading terms have been retained.

For the solar sectorial tides we must consider the product

$$(1 + 0.050 \cos l' + \cdots)(0.921 + 0.079 \cos 2h + \cdots) \cos 2t, \tag{60}$$

where $t = \tau + s - h$.

By using simple trigonometric identities we can immediately identify the angular arguments, and therefore the frequencies, of those sectorial tides (both lunar and solar) having the largest amplitudes. The arguments of the tides are given below; the Darwin symbol is listed alongside whenever available:

$$\begin{array}{llllll}
(1) & 2\tau & M_2 & (8) & 2\tau - 2D & \mu_2 \\
(2) & 2\tau + l & L_2 & (9) & 2\tau + 2D & \\
(3) & 2\tau - l & N_2 & (10) & 2t & S_2 \\
(4) & 2\tau + 2s & {}^{\mathrm{m}}K_2 & (11) & 2t - l' & T_2 \\
(5) & 2\tau - 2s & & (12) & 2t + l' & R_2 \\
(6) & 2\tau - l + 2D & \lambda_2 & (13) & 2t - 2h & \\
(7) & 2\tau + l - 2D & \nu_2 & (14) & 2t + 2h & {}^{\mathrm{s}}K_2
\end{array} \tag{61}$$

Since $t + h = \tau + s$, the tides ${}^{\mathrm{m}}K_2$ and ${}^{\mathrm{s}}K_2$ have the same frequency. The argument number of a solar tide can be easily ascertained by means of the relation $t = \tau + s - h$.

We can then express its argument in terms of the six fundamental variables. Thus, e.g., for the solar tide R_2, the argument can be written as

$$2t + l' = 2(\tau + s - h) + h - p_s = 2\tau + 2s - h + 0 \cdot p + 0 \cdot N_1 - p_s,$$

and this gives rise to the argument number 274.554. Note that it is customary to separate by a dot the first three digits of the argument number (referring to the short-period variables) from the last three digits (which refer to the long-period variables).

To consider the tesseral tides, we first express $\sin 2\delta$ according to the approximation

$$\begin{aligned} \sin 2\delta &= 2 \sin \delta \cos \delta \cong 2 \cos \delta \\ &= 2 \sin \epsilon \sin s = 0.79564 \sin s; \end{aligned}$$

next we expand the products

$$(0.795)(1.005 + 0.164 \cos l + \cdots) \sin s \cos \tau,$$
$$(0.795)(1 + 0.050 \cos l' + \cdots) \sin h \cos t.$$

We obtain the following tides:

(15)	$\tau - s$	O_1	(21)	$t - h$	P_1
(16)	$\tau + s$	${}^{m}K_1$	(22)	$t + h$	${}^{s}K_1$
(17)	$\tau + s + l$	J_1	(23)	$t + h + l'$	ψ_1
(18)	$\tau - s - l$	Q_1	(24)	$t + h - l'$	S_1
(19)	$\tau + s - l$	M_1	(25)	$t - h + l'$	
(20)	$\tau - s + l$		(26)	$t - h - l'$	π_1

(62)

Finally, for the zonal tides we must consider the expansions

$$-(1.005 + 0.164 \cos l + \cdots)(0.254 + 0.079 \cos 2s + \cdots),$$
$$-(1 + 0.050 \cos l' + \cdots)(0.254 + 0.079 \cos 2h + \cdots).$$

This yields the following tides:

(27)	0	M_0	(30)	0	S_0
(28)	$2s$	M_f	(31)	$2h$	S_{sa}
(29)	l	M_m	(32)	l'	S_a

(63)

Table 6.1 provides the fundamental tides, of both lunar and solar origin, that have an amplitude larger than $\frac{1}{2}H \cdot 10^{-3}$ and $\frac{1}{2}H' \cdot 10^{-3}$. It is obtained from the results of Cartwright and Tayler. The listing contains 118 terms due to the Ω_2 potential and 11 due to Ω_3. We give only the argument number and the amplitude of the tide.

The lunar and solar tides having the same argument—e.g., M_0, S_0; ${}^{m}K_2$, ${}^{s}K_2$; and ${}^{m}K_1$, ${}^{s}K_1$—cannot be distinguished from each other and are listed as combined lunisolar tides.

Table 6.1 Amplitudes of Lunar and Solar Tides Due to the Second- and Third-Order Tidal Potentials

Argument number	Amplitude	Argument number	Amplitude	Argument number	Amplitude	Argument number	Amplitude
055.555	0.738	125.755	0.009	167.555	−0.008	263.655	−0.007
055.565	−0.066	127.545	0.002	173.655	−0.006	265.455	−0.026
056.554	0.012	127.555	0.011	173.665	−0.001	265.655	0.006
057.555	0.073	135.645	0.014	175.455	−0.030	265.665	0.003
057.565	−0.002	135.655	0.072	175.465	−0.006	267.455	0.001
058.554	0.004	137.445	0.003	183.555	−0.005	271.557	0.001
063.645	−0.001	137.455	0.014	185.355	−0.002	272.556	0.025
063.655	0.016	143.755	−0.001	185.555	−0.016	273.555	0.422
063.665	−0.001	144.556	−0.001	185.565	−0.010	274.554	−0.004
065.445	−0.005	145.535	−0.002	185.575	−0.002	275.545	−0.001
065.455	0.083	145.545	0.071	195.455	−0.003	275.555	0.115
065.465	−0.005	145.555	0.377	195.465	−0.002	275.565	0.034
065.655	−0.004	145.755	−0.002	217.755	0.001	275.575	0.004
065.665	−0.002	146.554	0.001	225.855	0.003	283.655	0.001
067.455	−0.001	147.555	−0.005	227.655	0.007	285.455	0.006
073.555	0.014	147.565	0.001	229.455	0.001	285.465	0.003
075.355	0.007	153.655	−0.003	235.755	0.023	293.555	0.001
075.555	0.156	155.445	−0.002	237.545	−0.001	295.555	0.002
075.565	0.065	155.455	−0.011	237.555	0.028	295.565	0.001
075.575	0.006	155.655	−0.030	238.554	0.002		
083.455	0.002	155.665	−0.006	244.656	−0.001		
083.655	0.006	157.455	−0.006	245.645	−0.007	065.555	0.005
083.665	0.002	157.465	−0.001	245.655	0.174	135.555	−0.002
085.455	0.030	162.556	0.010	246.654	0.002	145.655	−0.001
085.465	0.012	163.545	−0.002	247.445	−0.001	155.555	−0.007
085.475	0.001	163.555	0.175	247.455	0.033	175.555	−0.002
093.555	0.005	164.554	−0.001	248.454	0.002	235.655	−0.002
093.565	0.002	164.556	−0.004	253.755	−0.003	245.555	−0.006
095.355	0.004	165.545	0.010	254.556	−0.003	265.555	0.005
095.365	0.002	165.555	−0.530	255.545	−0.034	345.655	−0.003
115.855	0.001	165.565	−0.072	255.555	0.908	355.555	−0.012
117.655	0.003	165.575	0.002	256.554	0.003	375.555	−0.002
125.745	0.002	166.554	−0.004	257.555	0.001		

6.07 EVALUATION OF THE GLOBAL OCEAN TIDES AS A BOUNDARY-VALUE PROBLEM

We resume our discussion of the LTE started in Section 6.01 using the same notation, but we consider now the disturbing potential Γ from a more sophisticated point of view. Γ will consist of (1) a certain astronomical harmonic Ω_m due to the Moon and Sun acting on both ocean and solid Earth; (2) a term due to the yielding of the solid Earth because of the fluid tidal layer; (3) a term representing the attraction of that tidal layer upon the rest of the ocean (i.e., ocean self-attraction); and (4) terms due to friction forces and viscosity effects.

We shall briefly examine these various constituents of the perturbing potential and show that the effect of the fluid tidal layer is to transform the original LTE into an integrodifferential equation.

Since the free periods of the elastic vibrations of the solid Earth are less than 1 hr and thus considerably shorter than the tidal periods, also due to high dissipation coefficients of the Earth's normal modes, we can safely express the response of the solid Earth to the nonloading potential Ω_m by means of the common Love numbers h_m, k_m, l_m and its response to the surface loading of the tidal column by means of the load numbers h'_m, k'_m, l'_m.

The response of the solid Earth to a given harmonic Ω_m of the potential due to an astronomical perturbing body is an uplift expressed by $h_m\Omega_m/g$; because of the ensuing variation in density distribution, we get an additive change $k_m\Omega_m$ to the original potential. The total potential is given by $(1 + k_m)\Omega_m$ and takes into account the attraction of the uplifted layer.

The tidally raised level of ocean water of density ρ_w and height ζ_0, measured from the uplifted sea bottom, will generate a normal load $\rho_w\zeta_0 g$ per unit surface area. This load will give rise to an elastic normal displacement of the solid Earth given by $u^*(\gamma)$, where γ is the angular separation of the location where the displacement is to be evaluated from the loading line. This displacement is due to the compression and attraction exerted by this additional layer of water on the solid Earth. We have shown in Chapter IV that for unit of mass loading, this deformation is expressed by

$$u^*(\gamma) = \frac{a}{m_1} \sum_{n=0}^{\infty} h'_n P_n(\cos\gamma), \tag{64}$$

where m_1 is the mass of the Earth. To account for the variable height of the loading column, one must integrate the above expression over the whole ocean surface; this gives rise to the elevation

$$U^*(\theta, \phi; t) = \iint_{\text{ocean}} \rho_w \zeta_0(\theta', \phi'; t) u^*(\gamma)\, d\Sigma'. \tag{65}$$

Here the integration variables (θ', ϕ') and the coordinates (θ, ϕ) of the position at which the integral is to be evaluated are related as follows:

$$\cos\gamma = \cos\theta\cos\theta' + \sin\theta\sin\theta'\cos(\phi - \phi'). \qquad (66)$$

This same liquid tidal layer will cause a change in potential given by

$$\psi^*(\gamma) = \frac{ag}{m_1}\sum_{n=0}^{\infty} k_n' P_n(\cos\gamma) \qquad (67)$$

per unit of mass loading. This expression must also be integrated for every point of the ocean surface; we get the contribution

$$\Psi^*(\theta, \phi; t) = \iint_{\text{ocean}} \rho_w \zeta_0(\theta', \phi'; t)\psi^*(\gamma)\, d\Sigma'. \qquad (68)$$

Finally, this tidal layer will attract the remaining ocean, causing a perturbation in the potential given by

$$W^*(\theta, \phi; t) = \iint_{\text{ocean}} \frac{\rho_w G \zeta_0}{2a\sin(\gamma/2)}\, d\Sigma'$$

$$= \iint_{\text{ocean}} \rho_w \zeta_0 \frac{ga}{m_1}\frac{d\Sigma'}{2\sin(\gamma/2)}, \qquad (69)$$

where G is, as usual, the universal gravitational constant. We can then write the elevation of the ocean bottom as

$$\delta = h_m\Omega_m/g + U^*(\theta, \phi; t), \qquad (70)$$

and the perturbing potential as

$$-\Gamma = g\bar{\zeta} = (1 + k_m)\Omega_m + W^*(\theta, \phi; t) + \Psi^*(\theta, \phi; t). \qquad (71)$$

Hence

$$\bar{\zeta} - \delta = (1 + k_m - h_m)\Omega_m/g + (1/g)(W^* + \Psi^*) - U^*. \qquad (72)$$

Let us finally mention that the dissipative forces (i.e., friction and viscosity), as they appear in the right-hand side of Eq. (4), are generally modeled as

$$F_f = -(C_f/D)(u^2 + v^2)^{1/2}\begin{bmatrix} u \\ v \end{bmatrix}, \qquad F_v = C_v\nabla^2\begin{bmatrix} u \\ v \end{bmatrix} \qquad (73)$$

in terms of the velocity components u, v. Here $D = h + \zeta_0$, ∇^2 is the Laplace operator, and $C_f \cong 0.003$ is a nondimensional parameter, whereas C_v has been assumed in various studies to be between 10^5 and 10^7 m^2/sec.

If we use Eq. (72) and express $\zeta - \bar{\zeta}$ as $(\zeta - \delta) - (\bar{\zeta} - \delta)$, we can write Eq. (4) as follows:

$$\begin{aligned}
\frac{\partial u}{\partial t} - 2\omega v \cos\theta &= -\frac{g}{a}\frac{\partial \zeta_0}{\partial \theta} + \frac{1 + k_{\mathrm{m}} - h_{\mathrm{m}}}{a}\frac{\partial \Omega_{\mathrm{m}}}{\partial \theta} \\
&\quad + \frac{1}{a}\frac{\partial \Gamma^*}{\partial \theta} - \frac{C_{\mathrm{f}}}{D}(u^2 + v^2)^{1/2} u + C_{\mathrm{v}} \nabla^2 u, \\
\frac{\partial v}{\partial t} + 2\omega u \cos\theta &= -\frac{g}{a \sin\theta}\frac{\partial \zeta_0}{\partial \phi} + \frac{1 + k_{\mathrm{m}} - h_{\mathrm{m}}}{a \sin\theta}\frac{\partial \Omega_{\mathrm{m}}}{\partial \phi} \\
&\quad + \frac{1}{a \sin\theta}\frac{\partial \Gamma^*}{\partial \phi} - \frac{C_{\mathrm{f}}}{D}(u^2 + v^2)^{1/2} v + C_{\mathrm{v}} \nabla^2 v,
\end{aligned} \tag{74}$$

where

$$\begin{aligned}
\Gamma^* &= W^* - \Psi^* - gU^* \\
&= \iint_{\text{ocean}} \zeta_0(\theta', \phi'; t)\rho_{\mathrm{w}} \left[\frac{ag}{m_1}\frac{1}{2\sin(\gamma/2)} + \psi^*(\gamma) - gu^*(\gamma)\right] d\Sigma'.
\end{aligned} \tag{75}$$

The third equation of the set, i.e., Eq. (5), remains the same:

$$\frac{\partial}{\partial \theta}(Du \sin\theta) + \frac{\partial}{\partial \phi}(Dv) = -a \sin\theta \frac{\partial \zeta_0}{\partial t}, \tag{76}$$

where $D = h + \zeta_0$.

The values of the Love and load numbers may be determined by previous considerations. Most often, only Ω_2 is considered and $\gamma_2 = 1 + k_2 - h_2$ is taken to be ~ 0.69.

Equations (74)–(76) constitute a system of integrodifferential equations in u, v, ζ_0 because ζ_0 appears in the integrand of Γ^*. Furthermore, because of the generality of the approach, we can account for the continental boundaries.

6.08 COMPUTATIONAL TECHNIQUES

The integrodifferential equations obtained in the previous section can be numerically solved by successive approximations using finite differences methods and the following boundary conditions imposed by an impermeable coastline: the component of the tidal velocity perpendicular to the coastline vanishes. Fundamental and pioneering work in this direction has been undertaken by Hansen (1962), Zahel (1973, 1977, 1978), Estes (1975, 1977), and Schwiderski (1978, 1979, 1980). The procedure consists essentially of

(1) solving the differential equations that result by neglecting the potential Γ^* to obtain a first-order approximation for u, v, ζ_0, (2) using these data to evaluate Γ^* as an integral, thus making it possible to begin the loop for solving the integrodifferential equation. The procedure should be repeated until convergence is reached; in this case this is equivalent to periodicity of the final solution and its independence of the initial state. It has been shown that the inclusion of the friction and viscosity forces produces a strong damping factor that contributes to rapid convergence. In this connection, we note that the previously named authors have found that for certain ocean regions convergence can be achieved within 5 tidal periods, whereas in other instances up to 16 iterations might be required.

The periodic tidal solutions are of the form

$$\zeta_0(\theta, \phi; t) = A(\theta, \phi)\cos[\sigma t + \beta(\theta, \phi)], \tag{77}$$

where the contours $A(\theta, \phi) = \text{const.}$ are called the corange lines and the family of curves $\beta(\theta, \phi) = \text{const.}$ are referred to as the cotidal lines. Corresponding to these periodic solutions, the expression for the potential Γ^* can be rewritten as

$$\Gamma^*(\theta, \phi; t) = P(\theta, \phi)\cos\sigma t - Q(\theta, \phi)\sin\sigma t, \tag{78}$$

where P and Q are integrals obtainable from Eq. (75) by replacing ζ_0 with $A\cos\beta$ and $A\sin\beta$, respectively.

An interesting feature that emerges from a numerical study of global ocean tides is the appearance of geographical regions where the tidal amplitude vanishes. These are called amphidromic regions and are the centers from which radiate cotidal lines. The direction according to which the tidal phase varies around an amphidromic region in the sense of progressively later rising high tides determines whether it is called a clockwise or a counterclockwise point. The most extensively studied families of tides have been the semidiurnal waves M_2, S_2, N_2, K_2 and the family of diurnal waves K_1, O_1, P_1.

The most commonly used computational techniques for the numerical evaluation of the integrodifferential equations of the tidal motion are central differences for the three space variables u, v, ζ_0, forward differences for the time variable, and a staggered grid where the velocity components u, v and the tidal elevation ζ_0 are computed at adjacent points. Coastal profiles are generally represented by horizontal lines passing through the u-points of southward velocity components and vertical lines passing through the v-points of eastward velocity components. This procedure facilitates the expression of the boundary conditions because the u-components of the velocity must vanish upon the horizontal shoreline and the v-components of the velocity must vanish along the vertical shoreline. Any coastal boundary can be most conveniently inputted by providing for each colatitude θ the longitudes of those points which constitute land–water boundaries. It is to be understood that a depth profile must be available and inputted in order to initiate computations.

Because of the singularity exhibited by the LTE at the North Pole ($\theta = 0$), most of the existing programs exclude the polar regions (say, latitudes larger than 80°) from computations. Computations for this situation can be made by considering the special analytic solution for the case of a constant depth ocean covering the whole globe and using it for the polar cap to match the numerical solution for higher colatitudes.

Any workable computer program should be able to treat computationally three matters. First, it has to be able to evaluate the elastic deformation $u^*(\gamma)$ and the increase in potential $\psi^*(\gamma)$ as infinite series of Legendre polynomials having the load numbers as coefficients. As discussed in Chapter IV, this can be achieved if we know the asymptotic values h^* and k^* of those two load numbers and use the Kummer summation method and a closed form expression for the sum of certain series of Legendre polynomials.

Second, it has to be able to evaluate the potential Γ^*, provided we can input the continental boundaries, a first-order approximation for the tidal elevation ζ_0, and the expressions for u^* and ψ^* in terms of the angular variable γ.

Third, it has to be able to operate recursively on the integrodifferential equation by inputting an initial solution and the potential Γ^*.

The use of finite differences will transform the first derivatives of u, v, ζ_0 into differences for these functions evaluated at neighboring grid points. Only the viscosity terms depending on the Laplacian will introduce second derivatives. These will give rise to values of the functions at three consecutive points; due to the presence of a^2 in the denominators in the Laplacian, these terms will be smaller than the others.

Most of the existing tidal maps are based on $6° \times 6°$ and $4° \times 4°$ grids. Only Zahel and Schwiderski have obtained maps on a $1° \times 1°$ grid.

It has been the consensus that the results obtained by the more refined model which includes the yielding of the solid Earth and the ocean self-attraction are more self-consistent and are in better agreement with the data gathered by direct measurements.

References

Adams, L. H., and Williamson, E. D. (1923). *J. Wash. Acad. Sci.* **13**, 418.

Alterman, Z., Jarosch, H., and Pekeris, C. L. (1959). *Proc. R. Soc. London, Ser. A* **252**, 80.

Alterman, Z., Jarosch, H., and Pekeris, C. L. (1961). *Geophys. J.* **4**, 219.

Bomford, G. (1971). "Geodesy," 3rd ed. Oxford Univ. Press, London and New York.

Boussinesq, J. (1885). "Applications des Potentiels à l'Étude de l'Équilibre et du Mouvement des SolidesÉlastiques." Gauthier-Villars, Paris.

Brown, E. W. (1905). *Mem. R. Astron. Soc.* **57**, 51.

Brown, E. W. (1919). "Tables of the Motion of the Moon," Vol. 3. Yale Univ. Press, New Haven, Connecticut.

Bullen, K. E. (1963). "An Introduction to the Theory of Seismology," 3rd ed. Cambridge Univ. Press, London and New York.

Cartwright, D. E., and Tayler, R. J. (1971). *Geophys. J. R. Astron. Soc.* **23**, 45.

Chandra, U. (1970). *Bull. Seismol. Soc. Am.* **60**, 639.

Darwin, G. H. (1883). *Br. Assoc. Adv. Sci., Rep.* **49–118**.

Darwin, G. H. (1886). *Proc. R. Soc. London* **41**, 337.

Darwin, G. H. (1907). "Scientific Papers," Vol. 1. Cambridge Univ. Press, London and New York.

Darwin, G. H. (1911). "Scientific Papers," Vol. 4, p. 201. Cambridge Univ. Press, London and New York.

De Sitter, W. (1924). *Bull. Astron. Inst. Neth.* **2**, 97.

Doodson, A. T. (1921). *Proc. R. Soc. London, Ser. A* **100**, 305.

Eckert, W. J., Jones, R., and Clark, H. K. (1954). "Improved Lunar Ephemeris 1952–1959," p. 283. US Govt. Printing Office, Washington, D.C.

Estes, R. H. (1975). "A Computer Software System for the Generation of Global Numerical Solutions of Diurnal and Semidiurnal Ocean Tides," BTS-TR-75-27.

Estes, R. H. (1977). "A Computer Software System for the Generation of Global Ocean Tides Including Self-Gravitation and Crustal Loading Effects," BTS-TR-77-82.

Farrell, W. E. (1972). *Rev. Geophys. Space Phys.* **10**, 761.

Ferrel, W. (1874). "Tidal Researches," U.S.C.G. and G.S. Report. Washington, D.C.

Frazer, R. A., Duncan, W. J., and Collar, A. R. (1938). "Elementary Matrices." Cambridge Univ. Press, London and New York.

Gilbert, F., and Dziewonski, A. M. (1975). *Philos. Trans. R. Soc. London, Ser. A* **278**, 187.

Haddon, R. A. W., and Bullen, K. E. (1969). *Phys. Earth Planet. Inter.* **2**, 35.

Hansen, W. (1962). "Hydrodynamical Methods Applied to Oceanographic Problems." Mitt. Inst. Meereskd., Hamburg University.

Hart, R. S., Anderson, D. L., and Kanamori, H. (1977). *J. Geophys. Res.* **82**, 1647.

Heiskanen, W. A., and Moritz, H. (1967). "Physical Geodesy." Freeman, San Francisco, California.

Hobson, E. W. (1955). "Spherical and Ellipsoidal Harmonics." Chelsea, New York.

Hough, S. S. (1897). *Philos. Trans. R. Soc. London, Ser. A* **189**, 201.

Hough, S. S. (1898). *Philos. Trans. R. Soc. London, Ser. A* **191**, 139.

James, R., and Kopal, Z. (1963). *Icarus* **1**, 442.

Jolley, L. B. W. (1961). "Summation of Series," 2nd rev. ed., No. 540. Dover, New York.

Kaula, W. M. (1966). "Theory of Satellite Geodesy." Ginn (Blaisdell), Boston, Massachusetts.

Knopp, K. (1951). "Theory and Application of Infinite Series." Blackie, London.

Kopal, Z. (1960). "Figures of Equilibrium of Celestial Bodies." Univ. of Wisconsin Press, Madison.

Kratzer, A., and Franz, W. (1960). "Transzendente Funktionen." Akad-Verlag, Leipzig.

Kuo, J. T., and Jachens, R. C. (1977). *Ann. Geophys.* **33**, 73.

Lamb, H. (1945). "Hydrodynamics," 6th ed. Dover, New York.

Landisman, M., Sato, Y., and Nafe, J. (1965). *Geophys. J. R. Astron. Soc.* **9**, 439.

Lanzano, P. (1962). *Icarus* **1**, 121.

Lanzano, P. (1969). *Astrophys. Space Sci.* **5**, 300.

Lanzano, P. (1971). *Astrophys. Space Sci.* **11**, 191.

Lanzano, P. (1974). *Astrophys. Space Sci.* **29**, 161.

Lanzano, P. (1975). *Astrophys. Space Sci.* **37**, 173.

Lanzano, P., and Daley, J. C. (1977). *AIAA J.* **15**, 1231.

Lanzano, P., and Daley, J. C. (1980). "Elastic Deformations of a Rotating Spheroidal Earth due to Surface Loads," NRL Rep. 8410.

Longman, I. M. (1962). *J. Geophys. Res.* **67**, 845.

Longman, I. M. (1963). *J. Geophys. Res.* **68**, 485.

Love, A. E. H. (1909). *Proc. R. Soc. London, Ser. A* **82**, 73.

Love, A. E. H. (1911). "Some Problems of Geodynamics." Dover, New York.

Melchior, P. (1978). "The Tides of the Planet Earth." Pergamon, Oxford.

Melchior, P., Moens, M., Ducarme, B., and Van Ruymbeke, M. (1981). *Phys. Earth Planet. Inter.* **25**, 71.

Mendiguren, J. A. (1973). *Geophys. J. R. Astron. Soc.* **33**, 281.

Moens, M. (1980). *Manuscr. Geodetica* **5**, 217.

Molodenskii, M. S. (1961). *Commun. Obs. R. Belg.* **288**, 25.

Molodenskii, M. S., Eremeev, V. F., and Yurkina, M. I. (1962). "Methods for the Study of the External Gravitational Field and Figure of the Earth" (translation from 1960 Russian edition). Israel Program for Scientific Translations, Jerusalem.

Munk, W. H., and Cartwright, D. E. (1966). *Philos. Trans. R. Soc. London, Ser. A* **259**, 533.

Munk, W. H., and MacDonald, G. J. F. (1960). "The Rotation of the Earth." Cambridge Univ. Press, London and New York.

Newcomb, S. (1960). "Spherical Astronomy." Dover, New York (first published in 1906 by Macmillan).

Pedlosky, J. (1979). "Geophysical Fluid Dynamics." Springer-Verlag, Berlin and New York.

Pekeris, C. L. (1966). *Geophys. J. R. Astron. Soc.* **11**, 85.

Pekeris, C. L., and Accad, Y. (1972). *Philos. Trans. R. Soc. London, Ser. A* **273**, 237.

Plummer, H. C. (1960). "An Introductory Treatise on Dynamical Astronomy." Dover, New York.

Poincaré, H. (1910). *Bull. Astron.* **27**, 321.

Schwiderski, E. W. (1978). "Global Ocean Tides," Part I, NSWC TR-3866.
Schwiderski, E. W. (1979). "Global Ocean Tides," Part II, NSWC TR 78-414.
Schwiderski, E. W. (1980). *Rev. Geophys. Space Phys.* **18**, 243.
Smart, W. M. (1953). "Celestial Mechanics," Longmans, Green, New York.
Smith, M. L. (1974). *Geophys. J. R. Astron. Soc.* **37**, 491.
Sokolnikoff, I. S. (1956). "Mathematical Theory of Elasticity," 2nd ed. McGraw-Hill, New York.
Struik, D. J. (1950). "Differential Geometry." Addison-Wesley, Cambridge, Massachusetts.
Tisserand, F. (1891). "Traité de Mécanique Céleste," Vol. II. Gauthier-Villars, Paris.
Varga, P. (1974). *Pure Appl. Geophys.* **112**, 777.
Wahr, J. M. (1981). *Geophys. J. R. Astron. Soc.* **64**, 677.
Watson, G. N. (1966). "Theory of Bessel Functions." Cambridge Univ. Press, London and New York.
Whittaker, E. T., and Watson, G. M. (1952). "Modern Analysis." Cambridge Univ. Press, London and New York.
Wiggins, R. A. (1968). *Phys. Earth Planet. Inter.* **1**, 201.
Woolard, E. W. (1953). "Theory of the Rotation of the Earth around its Center of Mass," Vol. XV, Part I. US Govt. Printing Office, Washington, D. C.
Zahel, W. (1973). *Pure Appl. Geophys.* **109**, 1819.
Zahel, W. (1977). *Ann. Geophys.* **33**, 31.
Zahel, W. (1978). *In* "Tidal Friction and Earth's Rotation" (P. Brosche and J. Sündermann, eds.), p. 98. Springer-Verlag, Berlin and New York.

Index

M

N

O

International Geophysics Series

EDITED BY

J. VAN MIEGHEM
(July 1959–July 1976)

ANTON L. HALES
(January 1972–December 1979)

WILLIAM L. DONN
Lamont-Doherty Geological Observatory
Columbia University
Palisades, New York

Volume 1 BENO GUTENBERG. Physics of the Earth's Interior. 1959

Volume 2 JOSEPH W. CHAMBERLAIN. Physics of the Aurora and Airglow. 1961

Volume 3 S. K. RUNCORN (ed.). Continental Drift. 1962

Volume 4 C. E. JUNGE. Air Chemistry and Radioactivity. 1963

Volume 5 ROBERT G. FLEAGLE AND JOOST A. BUSINGER. An Introduction to Atmospheric Physics. 1963

Volume 6 L. DUFOUR AND R. DEFAY. Thermodynamics of Clouds. 1963

Volume 7 H. U. ROLL. Physics of the Marine Atmosphere. 1965

Volume 8 RICHARD A. CRAIG. The Upper Atmosphere: Meteorology and Physics. 1965

Volume 9 WILLIS L. WEBB. Structure of the Stratosphere and Mesosphere. 1966

Volume 10 MICHELE CAPUTO. The Gravity Field of the Earth from Classical and Modern Methods. 1967

Volume 11 S. MATSUSHITA AND WALLACE H. CAMPBELL (eds.). Physics of Geomagnetic Phenomena. (In two volumes.) 1967

Volume 12 K. YA. KONDRATYEV. Radiation in the Atmosphere. 1969

Volume 13 E. PALMEN AND C. W. NEWTON. Atmospheric Circulation Systems: Their Structure and Physical Interpretation. 1969

Volume 14 HENRY RISHBETH AND OWEN K. GARRIOTT. Introduction to Ionospheric Physics. 1969

Volume 15 C. S. RAMAGE. Monsoon Meteorology. 1971

Volume 16 JAMES R. HOLTON. An Introduction to Dynamic Meteorology. 1972

Volume 17 K. C. YEH AND C. H. LIU. Theory of Ionospheric Waves. 1972

Volume 18 M. I. BUDYKO. Climate and Life. 1974

Volume 19 MELVIN E. STERN. Ocean Circulation Physics. 1975

Volume 20 J. A. JACOBS. The Earth's Core. 1975

Volume 21 DAVID H. MILLER. Water at the Surface of the Earth: An Introduction to Ecosystem Hydrodynamics. 1977

Volume 22 JOSEPH W. CHAMBERLAIN. Theory of Planetary Atmospheres: An Introduction to Their Physics and Chemistry. 1978

Volume 23 James R. Holton. Introduction to Dynamic Meteorology, Second Edition. 1979

Volume 24 Arnett S. Dennis. Weather Modification by Cloud Seeding. 1980

Volume 25 Robert G. Fleagle and Joost A. Businger. An Introduction to Atmospheric Physics. Second Edition. 1980

Volume 26 Kuo-Nan Liou. An Introduction to Atmospheric Radiation. 1980

Volume 27 David H. Miller. Energy at the Surface of the Earth. 1981

Volume 28 Helmut E. Landsberg. The Urban Climate. 1981

Volume 29 M. I. Budyko. The Earth's Climate: Past and Future. 1982

Volume 30 Adrian E. Gill. Atmosphere–Ocean Dynamics. 1982

Volume 31 Paolo Lanzano. Deformations of an Elastic Earth. 1982